Unlocking Potential: The Stem Cell Guide

Vani

TABLE OF CONTENTS

TABLE OF CONTENTS

TABLE OF CONTENTS

1

INTRODUCTION

1.1 STEM CELLS

1.1.1 Stem cells

Stem cells (SCs) are characterized by indefinite self-renewal capacity and the capability to differentiate into somatic cells with specialized functions (Hall & Watt, 1989). Depending on their differentiation potential, stem cells can be classified further as toti-, pluri-, multi- or unipotent. Totipotent SCs are capable to establish the formation of a complete organism/embryo including extraembryonic placental-cells and are only present from the zygote to the 8-cell morula state (Kelly, 1977). Pluripotent SCs, also referred to as embryonic SC (ESCs), can produce any cell or tissue of the three germlayer and germline lineage and physiologically reside in the inner cell mass of a blastocyst. The first isolation and generation of murine and human ESC-lines was successful over two decades ago providing an experimental platform to study SCs biology (Evans & Kaufman, 1981; Thomson et al., 1998). In addition to ESCs generated by isolation from zygotes, other approaches proved successful to obtain pluripotent SCs from adult cells. This includes the isolation of spermatogonial SCs, which turn into ESC-like cells during specific cell culture (Guan et al., 2006) and cell isolation from parthenogenic blastocysts, which are chemically induced oocytes (Robertson et al., 1983). Multipotent SCs are present postnatally and throughout adulthood. They have a narrower differentiation capacity into specific cell lineages than pluripotent SCs and mainly function to replace specific tissues in the adult organism. Examples of multipotent SCs include hematopoietic SCs that give rise to several blood cells (Ogawa, 1993) and mesenchymal SCs that can differentiate into osteoblasts, chondrocytes, myocyte and adipocytes (Dominici et al., 2006). Unipotent SCs are present with the most limited differentiation potential as they can only give rise to one cell type like spermatogonial SCs giving rise to sperm cells exclusively (Phillips et al., 2010).

2006 a "new" stem cell species was introduced: induced pluripotent stem cells (iPSCs). Yamanaka and Takahashi published in 2006 how already differentiated somatic cells can be reprogrammed back into a pluripotent state thereby reversing the one-way street of specialized somatic cells back into pluripotent stem cells (Takahashi & Yamanaka, 2006). Comparison studies of iPSCs with ESCs underscored high similarities in transcriptional profiles (Chin et al., 2009), epigenetic landscapes (Deng et al., 2009; Guenther et al., 2010) and differentiation capacities (Bilic & Izpisua Belmonte, 2012). This opened a new avenue for stem cell and biomedical research as individual-specific pluripotent stem cells became available circumventing the ethical issues in the utilization of *in vitro* generated embryos for the production of ESC-lines.

1.1.2 Reprogramming into iPSCs

In 2006 and 2007, two research groups described iPSC reprogramming in high ranking journals such as Cell and Science (Takahashi & Yamanaka, 2006; Yu et al., 2007) . In general, expression of defined gene transcripts need to be activated in a somatic cell in order to direct the cell back into a pluripotent state. Yamanaka and Takahashi were the first to publish their approach introducing the embryonic transcription factors (TF) Oct4, Sox2, c-Myc and Klf4 into mouse fibroblasts and demonstrated their reprogramming into iPSCs. These reprogramming factors became also known as OKSM or the Yamanaka factors in honor of their founder. They identified OKSM from a screen of 24 pluripotency-genes by sequentially eliminating one factor at a time (Takahashi & Yamanaka, 2006). In a follow-up study they demonstrated that reprogramming with OKSM is also applicable for human fibroblasts cell lines (Takahashi et al., 2007). In 2007, Thomson and colleagues denoted that human mesenchymal cells can be reprogrammed by introducing a different set of pluripotency TF: OCT4, SOX2, NANOG and LIN28, which is also referred to as OSNL factors (Yu et al., 2007). Both studies pointed out that reprogramming was impossible if either Oct4/OCT4 or Sox2/SOX2 was removed underscoring the importance of these two TF as master regulators. Notably, recent studies proposed that Oct4 is dispensable demonstrated by successful reprogramming of mouse fibroblasts with KSM (Klf-4, Sox2, c-Myc) (Velychko et al., 2019). Daley and colleagues finally demonstrated that human primary skin fibroblasts, isolated from a skin biopsy, can be reprogrammed by ectopic expression of OKSM thereby paving the way for a human stem cell system that can be generated from any individual (Park et al., 2008). Whether OCT4 can be omitted in the reprogramming of human material remains to be elucidated. The first reports of iPSC reprogramming used lenti- or retroviral transfection systems that integrated the reprogramming factors into the host genome. Ever since, many studies focused on enhancing reprogramming efficiency and factor delivery to streamline iPSCs generation and make them available for research and cell transplant therapies (Fig. 1A). Delivery of the OKSM by other viral systems like Adenovirus or Sendai virus hold the advantage of footprint-free reprogramming with reported efficiencies of 0.002 – 1.0 % (Zhou & Freed, 2009; Fusaki et al., 2009). In perspective to clinical application, a footprint-free non-viral reprogramming is favorable as residues from viral reprogramming could cause interference in clinical applications. For this, successful reprogramming with episomal expressing plasmids (Hu et al., 2011; Okita et al., 2011), mRNA (Warren et al., 2010) or proteins (Kim et al., 2009) has been reported to date with efficiencies that range from 0.02 to 1.4 %. Importantly, it was shown that the OKSM factors should not be used in equimolar amounts and that factor stoichiometry with for example higher OCT4 amounts increases reprogramming efficiency and quality (Tiemann et al., 2011). In addition to fibroblasts, successful reprogramming of other cell sources was shown for blood cells, keratinocytes and bone marrow or nasal mucosa mesenchymal cells (Hübscher et al., 2019; Streckfuss-Bömeke et al., 2013; Staerk et al., 2010).

This summarizes the tremendous effort undertaken to analyze and improve iPSC reprogramming producing multiple valuable insights and protocols to generate iPSCs.

1.1.3 Maintenance and differentiation of iPSCs

One major aspect in long-term culture is the maintenance of iPSCs in the undifferentiated state without exceeding genomic modifications. Formerly, iPSCs and ESCs were cultured with serum on feeder layer cells like mouse embryonic fibroblasts (Yue et al., 2012; Ito et al., 2007). This holds the disadvantage that animal-based products vary among batches and therefore manifest in inconsistent culture conditions. As the knowledge and research on iPSC increased, chemically defined media became available and improved culture conditions and the consistency for subsequent differentiation experiments. It has been shown that an array of small molecules is necessary to provide good iPSC proliferation, survival and stemness e.g. fibroblast growth factor to activate the tyrosine kinase-coupled FGF-receptors, LiCl for Wnt-pathway activation, TGF-β to activate serine/threonine kinases for SMAD phosphorylation, γ-amino butric acid, pipeocolic acid, L-Ascobic acid, and Insulin (Chen et al., 2011). IPSC-culture allows to study stem cell biology and in addition the pluripotency presents the option to differentiate in any given cell type providing the right developmental cues. In general, insights provided by embryogenesis studies are utilized to enhance or inhibit crucial signaling pathways to sequentially direct SCs into a differentiated cell of interest. Fortunately, many differentiation protocols formerly established for ESCs could be transferred to iPSCs differentiation and thereof many feasible protocols for iPSC-neurons, -oligodendrocytes, -astrocytes, -hepatocytes, -islet cells and -cardiomyocytes became rapidly available (Shi et al., 2017). As iPSC-cardiomyocytes (iPSC-CMs) represent an indispensable tool in the cardiac research field, they will be discussed further in the following sections.

A Reprogramming

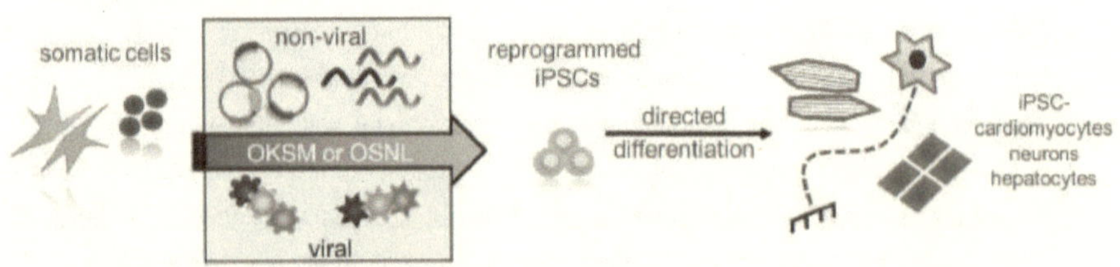

B Differentiation into iPSC-CMs

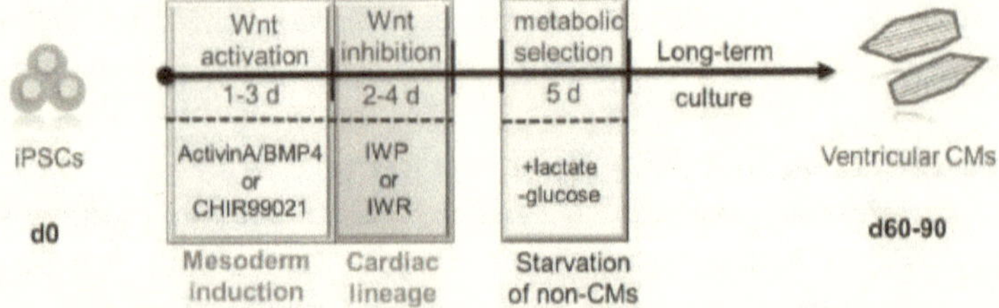

Fig. 1 Reprogramming and cardiac differentiation of iPSCs
A: Different approaches for reprogramming of somatic cells into iPSCs by non-viral (plasmid, mRNA) and viral vectors (Lenti- or Sendai virus). IPSCs can be differentiated subsequently into various cell types.
OKSM: OCT4, SOX2, KLF-4, c-MYC; OSNL: OCT4, SOX2, NANOG, LIN28
B: Differentiation into iPSC-CMs by Wnt pathway modulation. First, Wnt signaling is activated to induce mesoderm formation and subsequently inhibited to drive the iPSCs into cardiac progenitor cells. Prolonged culture of 60-90 d matures them into functional ventricular iPSC-CMs. To enhance CMs purity a metabolic selection with glucose-deprived medium can be included during culture.

1.1.3.1 Differentiation into iPSC-CMs

IPSC-CMs allow studying cardiac diseases in a human and even patient-specific context. This platform holds many advantages since iPSCs are less ethically burdened compared to ESCs, present the whole genetic background of a patient, and hold the potential to significantly reduce animal studies, especially in drug and toxicity screening applications. To generate iPSC-CMs, developmental cues from cardiogenesis are mimicked to direct the iPSCs into cardiac lineage commitment. During embryogenesis the heart is one of the first organs to develop starting in the 3rd week after fertilization (Moorman et al., 2003). Heart development requires a fine-tuned spatiotemporal activation of signaling pathways. The major pathways involved are nodal/bone-morphogenic protein (BMP) and Wnt signaling for generation of cardiac progenitor cells and NOTCH pathway signaling for chamber morphogenesis (Burridge et al., 2012). First, Wnt signaling is activated by BMP4 inducing mesodermal commitment. The cardiogenic area in an embryo is composed of mesodermal cells that will give rise to the epicardium and myocardium. The promotion towards cardiac lineage commitment requires BMP signaling while Wnt signaling is blocked giving rise to Mesp1+ cardiac progenitor cells. This shows that Wnt signaling has a biphasic effect in early cardiogenesis (Ueno et al., 2007). As the embryo folds, these cardiac progenitor cells are formed into a crescent-shaped heart tube divided into a primary heart field (PHF) and secondary heart field (SHF) area. The PHF is characterized by Nkx2.5, Tbx5 and Hand1 expression and will contribute to the left ventricle, atria and conduction system. The SHF is distinguished by expression of Isl1, Fgf10 and Hand2 and will subsequently form the right ventricle, atria and outflow tract (Buckingham et al., 2005; Srivastava, 2006; George & Firulli, 2019). One efficient approach to differentiate iPSCs into CMs is to mimic these developmental processes by sequential Wnt activation and subsequent Wnt inhibition. This is achieved by addition of physiological growth factors or small molecules. Starting with 2D monolayer or 3D embryoid body (EBs) iPSCs, Wnt activation for 1-3 d with activin A/BMP4 induces mesodermal lineage commitment. Alternatively, small molecules like CHIR99021 are used, which activate Wnt signaling by blocking the Wnt inhibitor glycogen synthase kinase 3. Subsequently, Wnt signaling inhibitor molecules like Inhibitor of WNT-production (IWPs) are added for 2-4 d to direct the cells into cardiac lineage. This protocol primarily produces ventricular iPSC-CMs, which can be enriched by metabolic selection with lactate to starve glucose-dependent non-CMs (Burridge & Zambidis, 2013; Lian et al., 2013; Tohyama et al., 2013) (Fig. 1B). The addition of retinoic acid (RA) during the second step (Wnt inhibition) is sufficient to produce another CM subtype, atrial iPSC-CMs (Cyganek et al., 2018). Sufficient derivation of pacemaker cells remains an ongoing challenge and protocols often resort to isolation of pacemaker cells from ventricular differentiation batches, ectopic overexpression of pacemaker specific ion channels like the hyperpolarization-activated cyclic nucleotide-gated cation channel (HCN) during cardiac differentiation or co-culture with visceral endoderm-like cells (Schweizer et al., 2017; Darche et al., 2019). In addition, non-CMs like cardiac fibroblasts and endothelial cells can be produced from iPSCs creating the opportunity to co-culture CMs and non-CMs in more physiological 3D structures (Zhang et al., 2019; Natividad-Diaz et al., 2019). This toolbox of cardiac relevant iPSC-differentiations provides a powerful tool as a human *in vitro* system to model and study cardiac diseases.

1.1.3.2 Maturation of iPSC-CMs

After directed differentiation, the iPSC-CMs exhibit an immature and fetal-like phenotype. Therefore, a big challenge is to mature iPSC-CMs since many cardiac diseases concern adult CM physiology. Many studies focused on the maturation state of iPSC-CMs compared to adult CMs and the possibilities to enhance maturation *in vitro*. Parameters to determine maturation status include gene marker expression, electrophysiological properties, cytoskeletal and sarcomere organization and metabolism. Similar to adult CMs, the iPSC-CMs express cardiac ion channels, sarcomeric proteins and Ca^{2+} handling genes, show autonomous beating and subtype-specific action potentials (Karakikes et al., 2015). However, iPSC-CMs show an overall lower expression of cardiac genes like ryanodine receptor 2 (*RYR2*), Na^+ voltage gated channel alpha subunit 5 (*SCN5A*) or K^+ voltage-gated channel subfamily H member 2 (*KCNH2*) (Yang et al., 2014a). Furthermore, they often lack gene isoform switches present in their adult counterparts, for example higher levels of fetal alpha-myosin heavy chain (*MYH6*) and titin-N2BA in iPSC-CMs instead of adult beta-myosin heavy chain (*MYH7*) and titin-N2B (Kolanowski et al., 2017; Ahmed et al., 2020). The morphology of healthy adult CMs is a rod shaped polyploid cell with a highly regular sarcomere structure and invaginations to the cell membrane called T-tubuli. IPSC-CMs also show a sarcomere structure and some polynucleated cells, however they are neonatal-like circular cells and often lack mature ultrastructural characteristics like T-tubuli or a mature sarcomere length of ca 2.2 µm. Furthermore, iPSC-CMs can only generate half the contractile force compared to an adult myocardium and exhibit a negative frequency-force relation (Lundy et al., 2013; Kamakura et al., 2013; Yang et al., 2014a; Machiraju & Greenway, 2019). Some of the decreased levels of cardiac gene transcripts can explain further immaturity features observed in their electrophysiology. The iPSC-CMs were described to show electrical activity and subtype-specific actions potentials (AP) were detected. However, compared to adult CMs, iPSC-CMs were described with higher resting membrane potential (RMP) (attributed to low expression of the channel for K⁺ inward rectifier current Kir2.1), slower upstroke velocity (attributed to low levels of *SCN5A*/Nav1.5) and decreased conduction velocities (attributed to low levels of connexins) (Lemoine et al., 2017; Horváth et al., 2018). Metabolically, adult CMs rely on fatty acid oxidation to satisfy the hearts energy demands. IPSC-CMs can metabolize glucose and lactate but lack a sufficient capacity for fatty acid metabolism (Correia et al., 2017). One widely used possibility for maturation is an increased culture time > 2 months, which yields increased cardiac gene expression and size, enhanced sarcomeric regularity and electrophysiological maturation (Kamakura et al., 2013; Lewandowski et al., 2018). Additional experimental approaches to enhance maturation include the following: culturing iPSC-CMs exclusively in glucose-free medium (Horikoshi et al., 2019), addition of small molecules like Tri-iodo-l-thyrodine (Yang et al., 2014b), applying electrical stimulation and mechanical stretch (LaBarge et al., 2019), co-culture with human mesenchymal stem cells (Yoshida et al., 2018) or culture in 3D structures or organoids (Tiburcy et al., 2017). To date, the optimal protocol to fully mature iPSC-CMs resembling all adult-like features remains to be determined.

1.1.3.3 iPSC-CMs in research, disease modelling and regenerative medicine

IPSCs can be generated from any individual and differentiation protocols allow to model and study different diseases in a tissue specific manner. Gene editing allows to generate specific isogenic control- and knockout lines or insertions of mutations into control-lines. In cardiac research, this is an invaluable tool to study monogenetic or multifactorial cardiac diseases in a human *in vitro* system. With regard to monogenetic disease variants, many studies underpinned the potential of patient-specific iPSC-CMs to recapitulate pathological phenotypes of dilative and hypertrophic cardiomyopathies and identified structural and/or electrophysiological impairments as an underlying disease phenotype (Karakikes et al., 2014). The possibility to generate isogenic control iPSC-lines further allows direct comparison and attributing a disease to single nucleotide polymorphisms (SNP). More complex and multifactorial cardiac diseases can also be studied, even if the genetic predisposition is unknown since patient-specific iPSCs harbor the disease specific genetics. For example, the Takotsubo- or Broken Heart syndrome is a complex temporary cardiomyopathy that affects mostly women post menopause. It was demonstrated, that iPSC-CMs from Takotsubo patients show higher sensitivity to β-adrenergic signaling as an underlying disease driver (Borchert et al., 2017).

In addition to disease modelling, iPSC-CMs are a promising player in the field of regenerative medicine. Loss of CMs due to myocardial infarction or injury results in the formation of non-CMs scar tissue. The iPSC-CMs are a potential source to replace this dysfunctional or diseased cardiac and scar tissue. Moreover, derivation of patient-specific iPSC-CMs can reduce immune response and rejection of transplants and avoids the need for immunosuppression therapy. Two major concerns for the application of iPSC-CMs are the formation of teratomas due to undifferentiated cells and the immaturity status of the cells that could lead to arrhythmias and graft rejection. However, preliminary studies in pig and monkey models showed beneficial effects of ESC/iPSC-CMs after myocardial infarction and first application tests in humans will follow within the next years (Romagnuolo et al., 2019; Shiba et al., 2016). Lastly, iPSC-CMs are applied in drug screening assays to discover new cardiac drugs or to test for cardiotoxic side-effects. Furthermore, screening of different medical compounds in patient-specific iPSC-CMs can help to identify the optimal medication strategy. This was demonstrated for an ACTN2-muation based hypertrophic cardiomyopathy, where medication of the patient was adjusted after Diltiazem showed better results in her iPSC-CMs (Prondzynski et al., 2019). In summary, iPSC-CMs are an invaluable human *in vitro* platform that is and will continue to be a crucial part in research and medicine.

1.1.4 Gene editing of iPSCs

The generation of iPSCs provides the opportunity to derive iPSCs from any healthy or disease-affected individual harboring their complete genetic background. In addition, gene editing by CRISPR/Cas9 further expands the iPSC platform by governing controlled genomic alterations for single nucleotide exchanges or gene knockdowns.

1.1.4.1 Clustered regulatory interspaced palindromic repeats/Cas9 technology

The clustered regulatory interspaced palindromic repeats (CRISPR)/Cas is a naturally occurring defense system against bacteriophages in bacteria and Archae (Grissa et al., 2007). It was discovered over two decades ago when researches identified distinct gene loci in *Escherichia coli* composed of arrangements of repeat and spacer segments with an associated protein coding segment (Ishino et al., 1987). CRISPR was chosen as a terminology to describe this genomic set-up consisting of repeat (23-50 nt) and spacer (17-84 nt) sequences with an abut protein coding gene <1kp downstream (Jansen et al., 2002; Haft et al., 2005) (Fig. 2A). The spacer sequences showed high variability and revealed sequence complementary to viral genomes. The repeat sequences are also very heterogeneous but share the potential to form stem loop structures. Analysis of the protein coding gene revealed an enzyme with two catalytic endonuclease activity domains (HNH and RuvC domains respectively) that could introduce double strand breaks (DSB), generally referred to as Cas proteins (Richter et al., 2012). Invasion of a bacteriophage triggers transcription of the CRISPR/Cas locus that arranges into an RNA/protein complex to specifically cleave and inactivate the invading viral genes. For this, the spacer and repeat segments transcribe into crispr RNAs (crRNA), complementary to parts of the viral genome and associate with the Cas protein via a trans-activating crRNA (tracrRNA). These crRNA serve as guides to direct the Cas protein to the specific site of action, where it will introduce numerous DSB to the invading viral genome. In addition to binding the target DNA via crRNA, the Cas proteins need another recognition site, termed protospacer adjacent motif (PAM), which is a 3 nt sequence adjacent to the protospacer/crRNA sequence. A typical PAM site is NGG, which is recognized by the Cas protein that will subsequently introduce a DSB 3 nt upstream of this PAM site. The use of PAM sites to fully activate the Cas protein is a protective mechanism that prevents cleavage of the crRNA (Sander & Joung, 2014). There have been numerous Cas proteins identified, distinguished by size and PAM recognition motif, which is used to classify CRISPR systems into 5 types (I-V). The type II system that is characterized by the use of Cas9 is the most frequently exploited system for genome editing in eukaryotic cells (Makarova et al., 2015).

In 2012, the first study described the use of CRISPR/Cas9 for as a tool for controlled gene editing (Jinek et al., 2012). In eukaryotic cells, DSB in the genome are recognized and repaired by two distinct mechanisms: non-homologous end joining (NHEJ) and homology directed repair (HDR). NHEJ is a fast but error-prone repair mechanism where the DNA ends, resulting from a DSB, are directly ligated back together (Pfeiffer & Vielmetter, 1988). HDR is a precise mechanism that uses a homologous piece of DNA as a template to reconstitute the DSB (Smithies et al., 1985). CRISPR/Cas9 technology is applied in eukaryotic cells to introduce site-specific DSB by designing the crRNA to bind the gene/position of interest. Subsequently, the DNA repair mechanisms will lead to random insertion/deletion mutations when the error-prone NHEJ mechanism repairs the DSB. This often leads to frameshift mutations with a premature stop codon that will disrupt the translational reading frame and thereof allows generation of targeted gene knockouts. In contrast, HDR relies on a template to repair the DSB typically provided by the other allele in the genome. However, HDR can also use an artificially designed DNA template that is provided exogenously. This artificial HDR template can be designed to induce controlled point mutations or to insert whole sequences to introduce e.g.

specific tags for the gene of interest (Sander & Joung, 2014) (Fig. 2B). To date, the modulation of the CRISPR system offers a wide range of gene editing possibilities in eukaryotic cells. Modification of the Cas9 protein leads to an expanded scope of applications by: A) Inactivation of one of the endonuclease domains (nCas), which will only introduce single-strand breaks. B) Complete inactivation of both endonuclease domains resulting in an unfunctional dCas protein. The dCas protein can be fused with transcriptional enhancer (e.g. VP64) or repressor domains (e.g. KRAB) and targeted to gene promotors for gene up- and downregulation, respectively. C) The dCas can also be fused to epigenetic modifiers to modulate gene expression pattern and chromatin activity. D) The dCas can be fused to a fluorescent protein, which allows to image genomic loci. E) Substitution of the Cas9 protein with other Cas proteins that recognize additional PAM sites or show decreased off-target cutting (Pickar-Oliver & Gersbach, 2019; Barrangou & Doudna, 2016; Kampmann, 2018).

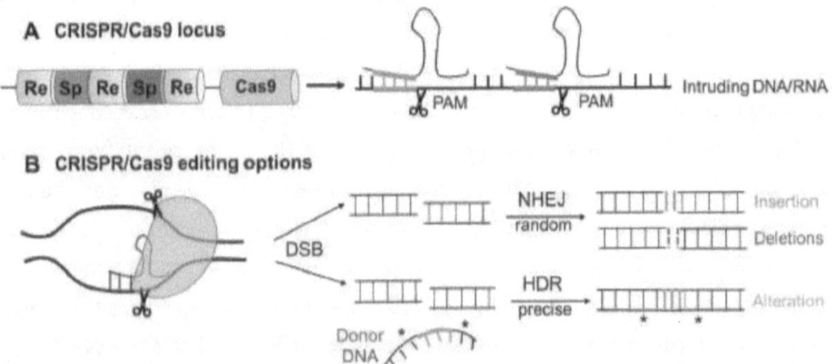

Fig. 2 CRISPR/Cas9 technology
A: Schematic representation of the bacterial CRISPR/Cas9 locus. Repeat (Re) and spacer (Sp) segments are arranged with an adjacent gene coding for the endonuclease Cas9 (orange). Re and Sp segments transcribe into RNA complexes (crRNA), whereas Re form a stem loop and Sp are complementary to viral DNA/RNA. The Sp part in the crRNA guide and bind the viral target and the associated Cas9 cleaves 3′ bp upstream of the protospacer adjacent motif (PAM) sequence thereby inactivating the viral intrusion.
B: Options for targeted gene editing approaches. Cas9 is guided by designed crRNAs to the gene of interest and introduces double strand breaks (DSB). These site specific DSB can either be repaired by non-homologous end joining (NHEJ) that leads to random insertion or deletions of nucleic acids or by homology directed repair (HDR) that uses a template DNA to remodel the DSB. For HDR, an exogenously provided HDR template allows to introduce specific nucleotide changes. * represents artificial silent mutations that are introduced to prevent re-cutting of the Cas9 after a successful edit.

1.1.4.2 CRISPR/Cas9 gene editing in iPSCs

Directed gene editing events in iPSCs can be achieved by CRISPR/Cas9 technology. Major difficulties faced are off-targets effects, low editing efficiency and CRISPR/Cas delivery into iPSCs. For CRISPR/Cas9 editing, two major components need to be brought into the cell: the guiding crRNA/tracrRNA complex and the Cas9 protein. Additionally, an HDR template is required if HDR-based point mutations or sequence insertions are desired. Delivery methods of CRISPR/Cas9 comprises coding plasmids, Adeno-associated virus (AAV) mediated transduction or electroporation of RNA (crRNA), ssDNA (HDR template) and protein (Cas9) directly (Lino et al., 2018). The latter has the advantage that use of protein will result in shorter Cas9 activity reducing off-target cutting compared to prolonged expression of Cas9 by plasmids. Furthermore, the use of Cas9 as a protein is less toxic for the application in stem cells (Okamoto et al., 2019). Delivery by AAV is a promising tool to apply and transport CRISPR/Cas9 into the human organism. However, the limiting packing capacity of AAVs surpasses the size of Cas9, which needs to be split if an AAV approach is chosen (Maeder et al., 2019). Despite this disadvantage, gene therapy using AAV delivery is investigated for medical applications due to its tissue tropism of the various AAV serotypes. These include AAV6 and AAV9 for gene delivery into cardiomyocytes (Ambrosi et al., 2019; Moretti et al., 2020), AAV5 for retinal cells (Maeder et al., 2019) and AAV2.5 (chimera o AAV2 and 5) for pulmonary epithelial cells (Li et al., 2019b). To further increase editing efficiency for directed single nucleotide exchanges, the HDR template is designed with silent mutations (leading to synonymous amino acids) for the crRNA or PAM site sequence to block further DSB events after a successful edit (Paquet et al., 2016). Reports of successful gene editing in iPSCs in biomedical research have been multitudinous. Functional gene knockouts in iPSCs by CRISPR/Cas9 using NHEJ enables to study mutation-based diseases in greater detail. Examples include the knockdown of the mutated allele *DNMT3B* to study the facial abnormalities syndrome (Horii et al., 2013) or *COL7A1* to study dystrophic epidermolysis bullosa (Shinkuma et al., 2016). Vice versa, this approach can also be applied to restore gene function exemplified by the excision of exon 45-55 in the *DMD* gene (Young et al., 2016). Many frameshift causing mutations in *DMD* lead to dysfunctional DMD and manifest in Duchenne muscular dystrophy. Deletion of exon 45-55 restored the translational reading frame and resulted in a shorter, but functional DMD (Young et al., 2016). Single nucleotide exchanges by CRISPR/Cas9 using an exogenously provided donor template for HDR allow to correct missense mutations and derive an isogenic iPSC-line from a patient iPSC-line. Inversely, a potential pathogenic SNP can be introduced into control iPSC-line to study the contribution of a single SNP to a certain disease (Ben Jehuda et al., 2018). The possibility to introduce controlled gene editing by CRISPR/Cas9 expands the power of iPSCs in research and disease modelling.

1.2 THE HEART: STRUCTURE, FUNCTION AND DISEASE

The heart represents the engine of our body by consistently pumping blood throughout the organism. The mammalian heart consists of two major and two minor chambers called ventricle and atrium, respectively. The ventricles fill with 60 - 120 ml blood and normally eject 50 – 75 % from their end-diastolic filling (Clay et al., 2006). The left ventricle supplies the body's circulation system and organs with oxygen and nutrient-rich blood. The smaller right ventricle extrudes blood to the lungs where it is released from carbon dioxide and loaded with oxygen. For sufficient heart performance, a fine-tuned interplay between multiple cellular mechanisms like electrical stimulation and propagation, Ca^{2+} cycling and structural features are crucial for the CMs to function as a beating force-generating syncytium.

1.2.1 Molecular architecture of the heart

For muscle cells to generate contraction, the CMs are composed of sarcomeres. Sarcomeres are the smallest contractual entity in CMs and numerous sarcomeres are arranged repetitively in one cell. The molecular architecture of a sarcomere is an ensemble of muscle specific proteins that are arranged transversely into thin and thick filaments. Optically, a sarcomere is segmented in the I-band (thin filaments adjacent to the Z-disc) and A-Band (overlapping thin and thick filaments) with the Z-disc as the lateral border and the M-line defining the center (Gautel & Djinović-Carugo, 2016; Clark et al., 2002) (Fig. 3A). The length of a sarcomere is measured between two Z-discs and is typically ~2.2 µm in human adult CMs (Freiburg et al., 2000). The major component of the thin filament is filamentous actin, which is anchored to the Z-disc by α-actinin. Nebulette proteins span the whole thin filament and serve as structural support for actin and its regulatory proteins tropomyosin and different troponins (cTNT, cTNC, cTNI). The primary protein of thick filaments is myosin, which is composed of myosin-heavy chains (MHC) and regulatory light chain (MLC) proteins. The giant protein titin (TTN) binds to the Z-disc with its N-terminal and to the M-line with its C-terminal domain thereby spanning half a sarcomere with one TTN protein providing structural support (Tajsharghi, 2008). The M-line is composed of M-protein, myomesin and obscurin and provides a scaffold to arrange the tick filaments into the A-band (Agarkova et al., 2004). Myosin from the thick and actin from the thin filaments interact and shift across each other pulling the Z-discs closer together in a multi-step sequence termed cross-bridge cycling thereby paving the foundation for contraction (Huxley, 1969). The cardiac contraction is triggered by Ca^{2+} influx, described further below. Structure, architecture, protein composition and arrangement of a sarcomere are crucial for functional contracting CMs. This becomes clear as many hereditary cardiac diseases show mutations in sarcomere proteins thereby disturbing the sarcomeric integrity and subsequently leading to reduction in heart performance.

1.2.2 A heartbeat: from membrane depolarization to Ca^{2+}-dependent contraction

CMs function as a syncytium and contract in a synchronized manner. The resting beat rate is generated from the pacemaker cells, travels via the sinoatrial node and finally propagates across the atria and ventricles. Such an electrical signal is converted into mechanical contraction via a delicate network of ion channels in the membrane, signaling proteins in the cytoplasm and finally sensing proteins in the sarcomere. First, the electrical stimulus will open voltage-gated Na^+ channels (Nav1.5) followed by Na^+ influx and membrane depolarization that will open voltage-sensitive L-type Ca^{2+} channels (Cav1.2) in the sarcolemma. This initial Ca^{2+} influx causes to open the Ca^{2+}- sensitive Ca^{2+} channels RYR2 in the sarcoplasmic reticulum (SR), which is termed Ca^{2+} induced Ca^{2+} release (CICR). Opening of RYR2 causes high Ca^{2+} efflux from the SR into the cytoplasm, which is sensed by the regulatory protein cTNT in the thin filament and initiates the crossbridge cycling (Lehman et al., 1994; Bers, 2008). To terminate contraction, ~70 % of Ca^{2+} is recycled back into the SR by sarcoplasmic/endoplasmic reticulum Ca^{2+} ATPase (SERCA) proteins and the remaining ions are recycled into the mitochondria by mitochondrial Ca^{2+} uniporter (MCU) and into the extracellular space by Na^+/Ca^{2+} exchanger (NCX) (Dedkova & Blatter, 2013) (Fig. 3B). Therefore, Ca^{2+} is the messenger that conveys excitation-contraction coupling and analysis of impaired Ca^{2+} cycling has been linked numerous times to dysfunctional and diseased myocardium (Kranias & Bers, 2007). During contraction, the cardiac AP represents the ion current events from different channels (Fig. 3C). The described Na^+ influx (I_{Na}) by Nav1.5 shapes the typical rapid depolarization in Phase 0 of the AP. Phase 1 is characterized by small repolarization caused by K^+ outward currents (I_{to}) from K^+ channels $K_v4.2$ and $K_v4.3$. The following Phase 2 plateau represents the Ca^{2+} influx (I_{ca}) events from Cav1.2 that will subsequently trigger the CICR. The rapid depolarization Phase 3 is by K^+ outward currents (I_{Kr}) from K^+ channels of the hERG family. The subsequent RMP (also referred to as Phase 4) is maintained by K^+ rectifier currents (I_{Ki}) from Kir2.1 channels, which is crucial to prepare the CMs for the next AP (Priest &

McDermott, 2015; Grant, 2009). Electrical activity in CMs relies on correct ion channel expression and function and multiple cardiac diseases with conduction or beating impairments have been linked to mutations in genes coding for ion channels collectively termed channelopathies (Behere & Weindling, 2015).

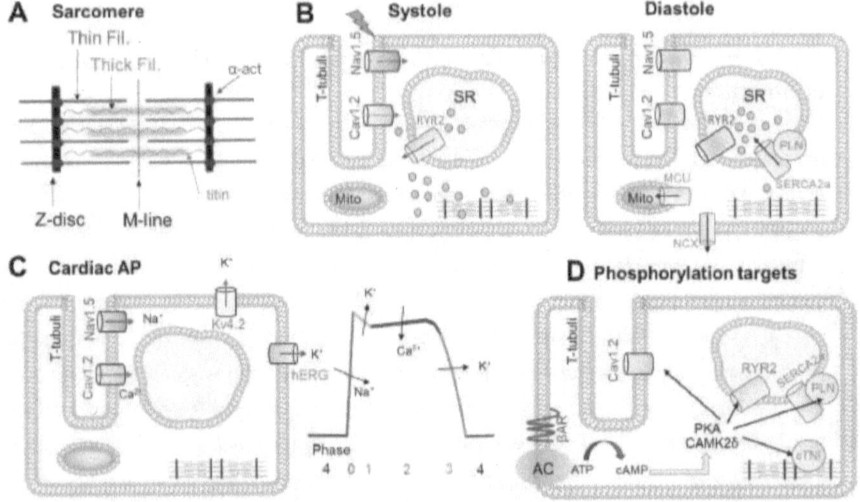

Fig. 3 Physiological properties of CMs
A: Schematic representation of a sarcomere with the transverse arranged thin (dark green) and thick (light green) filaments, the lateral Z-discs and the M-line. Fil. = filament. α-act (α-actinin) links the thin filaments to the Z-disc (purple). One titin molecule spans half a sarcomere to give structural support (red).
B: Ca^{2+} cycling in a CM and the involved ion channels. Opening of the Nav1.5 channels trigger opening of Cav1.2. Initial Ca^{2+} influx triggers opening of RYR2 (ryanodine receptor 2) releasing high amounts of Ca^{2+} from the SR resulting in contraction of the sarcomeres (Systole). Reuptake of Ca^{2+} terminates the contraction and leads to relaxation of the sarcomere (Diastole). SR = sarcoplasmic reticulum; MCU = mitochondrial Ca^{2+} uniporter, NCX = Na^+/K^+ exchanger; SERCA2a: ATPase sarcoplasmic/endoplasmic reticulum Ca^{2+} transporting 2
C: The typical cardiac AP is shaped by sequential openings of Na^+, Ca^{2+} and K^+ channels. AP = action potential.
D: β-adrenergic stimulation (βAR) activates phosphorylation by PKA and CAMK2δ of Cav1.2, RYR2, PLN and cTNI, which results in faster Ca^{2+} kinetics. AC = adenylate cyclase; PKA = protein kinase A

1.2.3 Faster and stronger: adrenaline and the cAMP cycle

The beat rate and contraction of the heart can be modulated by hormones for adjustment to physiological or environmental cues like exercise, stress or injury. One major player is the catecholamine adrenaline that has a positive chronotropic (beat rate↑), dromotropic (conduction speed↑), ionotropic (contractility↑) and bathmotropic (excitability↓) effect on the heart (Herman & Sandoval, 1983). Adrenaline conveys its effect by binding to G-protein coupled β-adrenergic receptors on the plasma membrane of CMs. This activates a signaling cascade that starts with the release of the G-protein from the receptor. The G-protein activates adenylate cyclases (AC), which produce cyclic adenosine monophosphate (cAMP) from intracellular adenosine triphosphate (ATP). The cAMP binds and activates protein kinase A (PKA) (Najafi et al., 2016). In parallel, CAMK2δ is activated, although the precise mechanism how β-adrenergic signaling conveys this effect is not fully understood (Grimm & Brown, 2010). Active PKA and CAMK2δ are functional kinases that phosphorylate Ser and Thr residues of their target proteins that comprise important Ca^{2+} handling proteins Cav1.2, RYR2, the SERCA regulatory protein Phospholamban (PLN) and the sarcomeric cTNI (Landstrom et al., 2017) (Fig. 3D). The phosphorylation events by PKA and CAMK2δ convey the described effects of adrenaline including the following: A) Phosphorylation of Cav1.2 and RYR2 enhances Ca^{2+} release and thereby increases muscle contractility. B) Phosphorylation of PLN decreases the inhibitory effect on SERCA and lead to enhanced Ca^{2+} uptake in the SR and muscle relaxation. C) Phosphorylation of cTNI reduces Ca^{2+} sensitivity in the myofilaments resulting in faster Ca^{2+} dissociation (van der Velden & Stienen, 2019).

1.2.4 Diseases of the heart: Cardiomyopathies

Cardiovascular diseases remain the world leading cause of mortality accounting for 1 in 4 deaths (Lozano et al., 2012). The entity of cardiomyopathies, which comprises all cardiac diseases that concern the heart muscle/myocardium in the absence of coronary artery or valve diseases. Cardiomyopathies can be classified by their physical manifestation into hypertrophic cardiomyopathy, dilated cardiomyopathy (DCM), arrhythmogenic right ventricular cardiomyopathy, restrictive cardiomyopathy, and left ventricular non-compaction (LVNC) as a subtype of unclassified cardiomyopathies (Elliott et al., 2008). LVNC and DCM will be described more detailed in the following sections.

1.2.4.1 DCM: a common cardiomyopathy

DCM accounts for 50-60 % of all cardiomyopathies and is characterized by an enlarged left ventricle with simultaneously reduced cardiac muscle (Albakri, 2018) (Fig. 4A). In patients, this can lead to many functional aberrations like systolic or diastolic dysfunction, poor contractile force, reduced ejection fraction, arrhythmias and in severe cases heart failure. End-stage heart failure caused by DCM accounts for half of heart transplantations (Regitz-Zagrosek et al., 2010). DCM can be caused by genetic and non-genetic factors. Non-genetic factors to acquire a DCM include hypertension, toxins, drug use, inflammation, or infectious diseases. Genetic factors or predisposing mutations were identified by screening families with hereditary heart diseases including multiple affected related family members and generations. It is estimated that 25-30 % of DCM cases are hereditary with many gene variants identified as disease causing. However, to date 50 % of DCM cases are still classified as idiopathic, meaning no primary or secondary cause of DCM can be identified (Albakri, 2018; McKenna et al., 2017).

1.2.4.2 LVNC: a rare cardiomyopathy

LVNC is a rare developmental cardiomyopathy with an estimated diagnosis of 0.1 – 1.3 % of patients undergoing echocardiograms. Patients are diagnosed with LVNC if their endocardial myocardium shows a non-compacted layer with hypertrabeculation and deep intertrabecular recesses (Jenni et al., 2007) (Fig. 4A). LVNC is considered a genetic and developmental heart disease, because it develops due to insufficient myocardial compaction and/or incomplete trabeculation reduction during heart development leaving the patient with the characteristic "spongy" and non-compacted myocardial wall (Wengrofsky et al., 2019). LVNC is rare and affected patients often present with similar symptoms to DCM like reduced ejection fraction and heart performance as well as atrial and ventricular arrhythmias (Arbustini et al., 2016). The definition of LVNC by the European Society of Cardiology categorizes LVNC as a standalone cardiomyopathy (Elliott et al., 2008) while the American Heart Association counts LVNC as a subtype of DCM (Maron et al., 2006). This underscores that little is known about the pathogenesis of LVNC and more research is necessary to elucidate the developmental defect and underlying molecular and physiological pathologies to better understand LVNC.

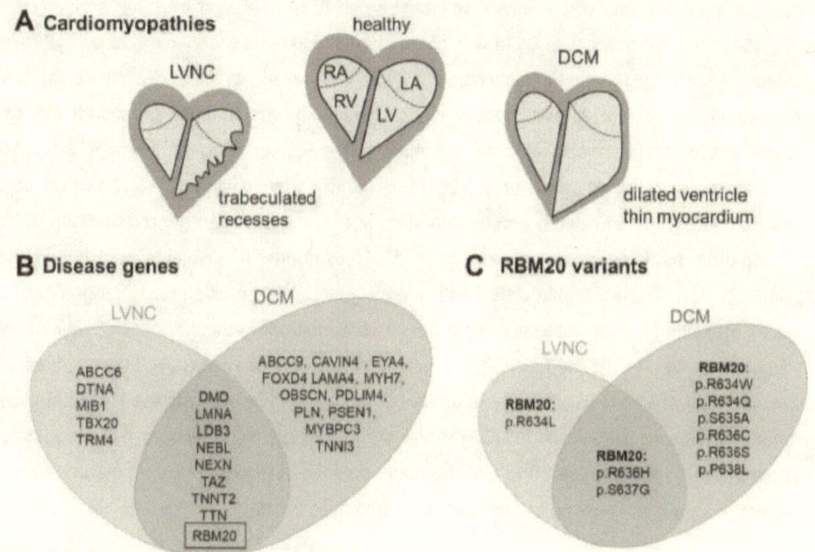

Fig. 4 Cardiomyopathies: LVNC and DCM
A: Morphological characterization of LVNC and DCM. LVNC shows deep trabecular recesses in the myocardial wall. DCM shows a dilated ventricle and thinning myocardium. RA: right atrium; RV: right ventricle; LA: left atrium; LV: left ventricle
B: Genes that are associated with genetic LVNC and DCM. Gene mutations summarized from (Brauch et al., 2009; Karakikes et al., 2014; Sedaghat-Hamedani et al., 2017; McNally & Mestroni, 2017; Miszalski-Jamka et al., 2017).
C: Different disease-causing variants of RBM20 associated with LVNC and DCM. RBM20 mutations summarized from (Brauch et al., 2009; Li et al., 2010; Sedaghat-Hamedani et al., 2017; Sun et al., 2020)

1.2.4.3 Genetics behind DCM and LVNC

Around 30 % of DCM patients have a genetic predisposition and it is estimated that up to 50 % of idiopathic DCMs have a genetic cause that is yet to be identified. To date, around 30 genes have been linked to carry disease-associated variants/mutations (Rosenbaum et al., 2020; McNally & Mestroni, 2017). Most of these genes code for important structural and electrophysiological cardiac proteins. Genetic mutations in sarcomeric genes like *TTN*, *MYH7*, and *cTNT* are associated to the development of DCM, mostly without any conduction impairment (Kamisago et al., 2000; Herman et al., 2012; Hershberger et al., 2013). In contrast, genetic variants in the nuclear envelope gene LaminA/C (*LMNA*) or the cardiac sodium channel (*SCN5A*) are associated with DCM with conduction diseases like tachycardias, arrhythmias, or sinoatrial node dysfunction (McNair et al., 2004; Cowan et al., 2018; Peretto et al., 2019). Additionally, there were also mutations identified in genes that do not contribute to the sarcomere or ion channels like laminin-alpha 4 or the RNA-binding motif protein 20 (*RBM20*) (Brauch et al., 2009; Abdallah et al., 2019). Genetic predispositions of LVNC remain poorly understood. A well correlated genetic mutation affects the Tafazzin gene that leads to LVNC among other pathologies summarized as the Barth syndrome (Ronvelia et al., 2012; Finsterer, 2019). In Fig. 4B genes associated with LVNC and DCM are illustrated. Intriguingly, some genes are linked to both diseases depending on the mutation within the gene. One example is RBM20 that causes mainly DCM, however one particular mutation (p.R634L) is linked to the development of LVNC (Sedaghat-Hamedani et al., 2017). Fig. 4C lists the RBM20 missense mutations that are associated to DCM and/or LVNC. Genetic screenings of patient cohorts are important in ongoing research to unravel novel disease-causing gene variants and to stratify patient groups with known gene mutations for optimal treatment.

1.3 RNA-SPLICING

Splicing describes the process by which intronic or exonic sequences are removed from a pre-mRNA transcript. It is a crucial part of the RNA maturation process before translation. This genetic architecture of introns and exons expands the proteomic diversity as multiple transcripts can arise from a single gene locus resulting in multiple protein isoforms. This becomes evident by the estimated 60.000 proteins identified in the human organism that originate from only 20.000 known protein-coding genes (Barbosa-Morais et al., 2012).

1.3.1 The spliceosome

Almost 99 % of human genes are composed of protein-coding sequences (exons) and non protein-coding sequences (introns) and are therefore subjected to splicing before translation (Barbosa-Morais et al., 2012). In human cells, the removal of introns is catalyzed by a specialized complex termed spliceosome. This spliceosome contains multiple small-nuclear RNA-protein complexes termed small nuclear ribonucleotide particles (snRNP) that are numbered as U1-, U2, U4-, U5- and U6- snRNP and over 100 additional proteins (Hastings &

Krainer, 2001; Rappsilber et al., 2002). The splicing of genes is intricate and orchestrated by a number of *cis*- acting cues within the gene (Fig. 5A). *Cis*-acting cues are encoded in the DNA of the genes itself. These so-called core splice sites are composed of a 5′ splice site containing a GU at the exon-intron border, a 3′ splice site with AG at the intron-exon border with a preceding polypyrimidine stretch and a branch point sequence (Zhang, 1998; Hubé & Francastel, 2015) (Fig. 5A). The removal of an intron occurs in two catalytic transesterification steps for which the spliceosome undergoes stepwise assembly and dynamic rearrangements. First, the 5′ splice site is bound by U1-snRNP, the branch point sequence by U2-snRNP and the polypyrimidine stretch and 3′ splice site by the U2-auxiliary factor heterodimer (U2AF2/U2AF1) collectively termed spliceosome complex A. Next, the U4/6 and U5 snRNPs are recruited to complex A followed by the detachment of U1 and U4 snRNPs resulting in the formation of the catalytic active spliceosome complex C (Lee & Rio, 2015; Will & Lührmann, 2011). Here, the reactions to cleave, remove the intron and join the exons are catalyzed. The first transesterification reaction is a nucleophilic attack from the branch point adenosine (A) at the 5′ splice site forming an intron lariat intermediate. The second transesterification reaction is a subsequent nucleophilic attack from the 5′ splice site at the 3′ splice site resulting in the ligated exons and a released intron lariat product (Lee & Rio, 2015) (Fig. 5B). Finally, the spliceosome detaches and is recycled for further splicing. Notably, 96 % of all introns are subjected to splicing in that complex, but there is also a small subset of introns removed by a so-called minor spliceosome or self-splicing events (Turunen et al., 2013), which will not be discussed in detail here.

1.3.2 Alternative splicing

The known discrepancy between expressed proteins (60.000) and protein-coding genes (20.000) is due to alternative splicing. This process explains how different peptides and proteins can arise from a single gene thereby increasing the coding capacity of the genome. Intriguingly, it was shown that almost all intron-containing genes are also subjected to alternative splicing (Pan et al., 2008). Splicing is not restricted to exclusively removing introns from a transcript. In addition, exons or parts of an exon can also be subjected to splicing, which results in multiple isoforms arising from the same gene locus. Whether an exon or intron is included or excluded in alternative splicing is controlled by additional *cis*- and *trans*-acting elements guiding the spliceosome to the site of action. In addition to the core splice sites described above, regulatory silencer or enhancer motifs are encoded in exonic or intronic segments of a gene (Wang & Burge, 2008) (Fig. 5A). These regulatory sequences are recognized and bound by a variety of splice factors that govern the activation or

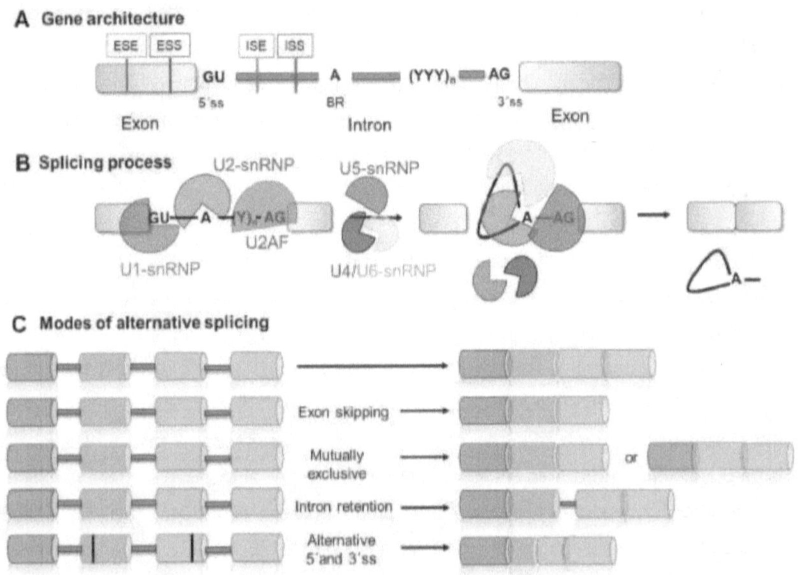

Fig. 5 Splicing mechanisms
A: Eukaryotic gene architecture harbors cues to splicing. The core splice sequences are an GU at the 5´ss, a branch point A, 3´pyrimidine stretch (YYY)n and an AG at the 3´ss. Additional cues are silencing and enhancer elements located in exonic and intronic regions. Ss = splice site; ESE: exonic splice enhancer; ESS: exonic splice silencer; ISE: intronic splice enhancer; ISS: intronic splice silencer
B: Splicing process conveyed by the spliceosome. First, recognition of the exon borders by binding of U1-snRNP, U2-snRNP and U2AF to the core splice sites. Next, assembly of an catalytically active spliceosome results in ligation of the exons and release of the intron in a lariat structure.
C: Modes of alternative splicing. Alternative splicing can produce multiple transcripts of varying length and usage of exonic and intronic elements.

repression of splicing by influencing splice site selection and activity (Solis et al., 2008). Two main classes of splice factors can be distinguished: heterogenous nuclear ribonucleoproteins (hnRNP) and SR proteins. As the name implies, hnRNP splice factors show structural heterogenicity and contain one or two RNA-recognition motifs (RRM) and one or two protein domains for protein-protein interaction (Geuens et al., 2016). SR proteins summarizes proteins with a conserved structure defined by a protein domain of repetitive Arg and Ser (RS) residues.

Canonical SR splice factors contain an N-terminal RRM for RNA binding and the RS domain in the C-terminal region, which functions for protein-protein interactions (Zhou & Fu, 2013). Both, hnRNPs and SR proteins, can bind to exonic and intronic regulatory segments of a gene where they function as splicing repressors or activators underscoring the diverse function of splice factors in alternative splicing (Lee & Rio, 2015). With this wide repertoire of regulation, multiple possibilities arise: exon skipping, intron retention, mutually exclusive exons and usage of alternative splice sites (Fig. 5C).

1.3.3 Splicing in disease

Splicing is a highly regulated and dynamic process to which the vast majority of human genes is subjected. Therefore, it is not surprising that an estimated 20 % of diseases with an underlying genetic cause are attributed to faulty splicing and a variety of genetic splice-associated diseases have been reported to date (Krawczak et al., 2007; Lim et al., 2011). Most common are *cis*-acting mutations that affect the splice site or regulatory splice elements within an individual gene that subsequently lead to missplicing and functional impairment of the affected gene. Mutations within the core splice sites of a gene often leads to exon skipping or intron retention resulting in reading frame shifts. These frame shifts often include the insertion of a premature stop codon, which is accompanied by nonsense-mediated decay (NMD) of the shortened transcript and thereby a functional loss of function. A prominent example are various splice site mutations in the *DMD* gene leading to exon skipping and subsequently NMD. This causes impaired function and levels of DMD that leads to progressive muscle wasting known as Duchenne muscular dystrophy (Juan-Mateu et al., 2013). In addition, pathological SNPs can be located in the regulatory exonic or intronic enhancer or inhibitor sequences of a gene mostly leading to an altered exon inclusion-exclusion ratio of the respective transcript. For example, this was reported for a mutation in the *MAPT* gene, which encodes the neuronal tau protein that is a major player in neurodegenerative diseases. Mutations in the regulatory elements of intron 9 result in increased inclusion of exon 10 and altered tau function and are linked to familial forms of frontotemporal dementia and Parkinsonism linked to chromosome 17 disease (Malkani et al., 2006; Faustino & Cooper, 2003).

Trans-acting mutations affect components of the spliceosome or regulatory splice factors. For example, many mutations in the TAR DNA-binding protein-43 (TDP-43) are linked to familial cases of amyotrophic lateral sclerosis (ALS), a neurodegenerative disease progressing into paralysis. It was demonstrated that TDP-43 binds to intronic sequences and is involved in splicing events of many genes for neurological function like MEF2D, parkin, huntington and ataxin. Coincidently, missplicing in these genes is observed in TDP-43-mutation carriers and benefits ALS disease progression (Da Cruz & Cleveland, 2011). Finally, misregulated RNA-splicing was also demonstrated as a driver in cardiac diseases highlighted by the discovery of RBM20-mutation based cardiomyopathies (Fig. 4C), which will be discussed in more detail below.

1.3.4 RBM20: a cardiac splice factor involved in cardiac disease

RBM20 is a splice factor highly expressed in the heart. Multiple patient cohort screenings revealed RBM20 variants as disease causing for DCM and LVNC (Brauch et al., 2009; Li et al., 2010; Sedaghat-Hamedani et al., 2017; Sun et al., 2020). These findings were complemented by Guo and colleagues, who identified RBM20 as a cardiac splice factor for *TTN*. They observed a larger *TTN* isoform N2BA expressed in RBM20 knockout rats and could link this longer *TTN* isoform to increased sarcomeric elasticity and myocardial stiffness (Guo et al., 2012). These pathologies could be confirmed later by Streckfuss et al., for a human RBM20 disease variant RBM20-p.S635A that recapitulated *TTN* missplicing and myocardial stiffness in a patient-specific iPSC-model (Streckfuss-Bömeke et al., 2017). In addition to *TTN*, 30 more RBM20 splice targets were identified to date, all of which encode proteins crucial for cardiac function (Maatz et al., 2014; van den Hoogenhof et al., 2018; Guo et al., 2012). RBM20 is a tissue-specific SR protein with two Zn-finger domains, one N-terminal RRM and one highly conserved C-terminal RS domain (Guo et al., 2012) (Fig. 6A). The RRM is important for binding of the RNA and a study by Maatz and colleagues unveiled direct binding of RBM20 to spliceosomal U1- and U2-snRNPs and to its target genes like *TTN*. They showed that RBM20 predominantly binds intronic sequences identifying UCUU as a core consensus sequence binding sequence. It was proposed that RBM20 mainly functions as a repressor of splicing by stalling of spliceosome complex A (Maatz et al., 2014). However, it was also reported that RBM20 functions in other modes of alternative splicing including exon exclusion, mutually exclusive exons and intron retention underscoring the complex role of RBM20 in splicing (Guo et al., 2012; Maatz et al., 2014) (Fig. 6B). The identification of human RBM20 variants unveiled the RS domain as a hot spot for disease causing missense mutations (Li et al., 2010; Brauch et al., 2009). The RS domain is characteristic for SR proteins and is termed after the repetitive sequence of Arg and Ser and is highly conserved among species from zebrafish to human. The RS domain of RBM20 (as well as the RS domain of splice factors in general) was shown to serve two major functions. First, Arg and Ser contain OH-groups and can be phosphorylated. These RS-phosphorylations have been shown to serve as localization cues, especially nuclear retention signals in RBM20 and other splice factors (Cazalla et al., 2002; Filippello et al., 2013).

Second, RS domains are crucial for protein-protein interactions, especially for interacting processes within the spliceosome (Shen & Green, 2006). In addition, splice factors often compete for splice sites to regulate alternative splicing. For RBM20, the polypyridine tract binding protein 4, an hnRNP splice factor, was shown to antagonize RBM20 splicing in *TTN* adding another layer of splicing control (Dauksaite & Gotthardt, 2018). In summary, RBM20 is an essential splice factor to regulate alternative splicing in > 30 cardiac genes. Many RBM20 disease variants have been identified, however the molecular and cellular mechanisms that lead to human cardiac diseases remain to be fully elucidated.

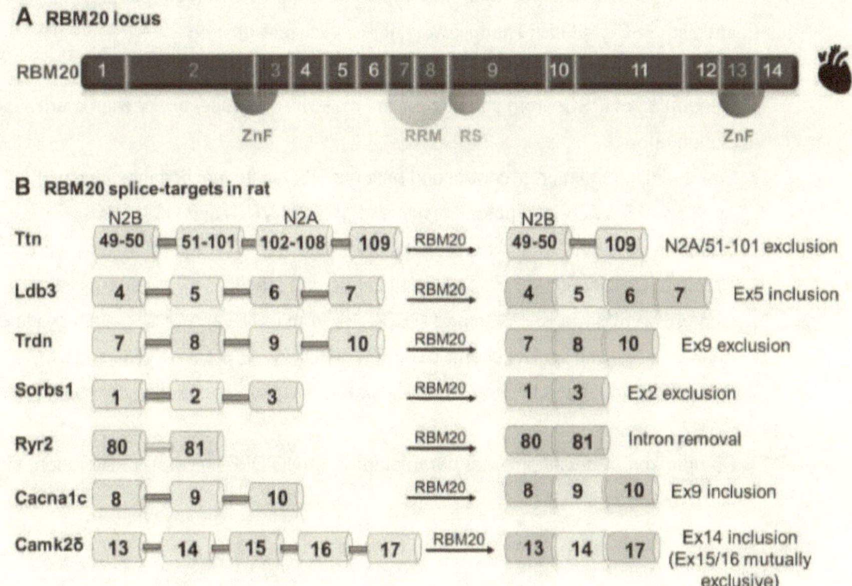

Fig. 6 RBM20 splice-targets
A: The RBM20 gene locus. RBM20 is composed of 14 exons with highly conserved domains: ZnF: Zinc finger; RRM: RNA recognition motif; RS: repetitive Arg/Ser domain
B: Schematic of RBM20 splice-targets based on rat model. Splicing represents wt-RBM20 function. Summarized from (Guo et al., 2012; Maatz et al., 2014; Watanabe et al., 2018). Ex = exon

1.4 AIMS

RBM20 is a cardiac splice factor that orchestrates alternative splicing of > 30 cardiac genes. Multiple RBM20 mutations that cluster in the RS domain have been linked to familial cases of cardiomyopathies. Two distinct heterozygous missense mutations of RBM20 concerning the p.R634 position have been shown to result in two different cardiomyopathies, namely p.R634L (LVNC) and p.R634W (DCM). The aim of this book was to investigate if and how these two RBM20 disease variants can result in different disease phenotypes using a patient-specific iPSC-platform. The following objectives were defined:

1. Generation of iPSCs from somatic patient material and subsequent characterization of pluripotency.
2. Directed differentiation of control and patients iPSCs into functional beating iPSC-CMs.
3. Analysis of RBM20- and potential splice targets in LVNC- and DCM-CMs.
4. Comparison of cellular and functional parameters in LVNC- and DCM-CMs: sarcomere structure, Ca^{2+} kinetics, Ca^{2+} load, beating rate and reaction to β-adrenergic stimulation.
5. Generation of isogenic iPSCs and iPSC-CMs from LVNC and DCM patient cell-lines.
6. Analysis of cellular and functional parameters of isogenic rescue LVNC and DCM iPSC-CMs and comparison to disease iPSC-CMs to assess the pathological contribution of RBM20 variants.
7. Comparison of developmental parameters in LVNC-CMs: growth, proliferation, single cell expression profiles.

2

MATERIAL

2.1 CHEMICALS, REAGENTS and ENZYMES

Tab. 1 List of chemicals, reagents and enzymes

	Company	Order number
Agarose powder	VWR	35-1020
Amino caproid acid	Applichem	A2266,0500
Biocoll	VWR	L6113
Boric Acid	Roth	6943.1
BSA powder Albumin Fraction V	Roth	8076.3
CaCl$_2$ solution	Sigma-Aldrich	21115
Caffeine powder	Sigma-Aldrich	C0750
DMSO	Sigma-Aldrich	D2650
dNTP mix (100 mM)	Bioline	Bio-39029
DTT	Roth	6908
Dynabeads M280	Thermo Fisher Scientific	00810740
Ethanol	Chemsolute	2236.1000
Fluo-4 probe	Thermo Fisher Scientific	F14201
Fluoromount-G	Southern Biotech	0100-01
FURA-2-am probe	Thermo Fisher Scientific	F1221
Gene ruler 100 bp	Thermo Fisher Scientific	SM0241
GoTaq G2 Polymerase	Promega	M784B
Green buffer 5x	Promega	M7911
HEPES	Roth	9105
Hydrochloric acid 37 %	Merck	1.00314.1000
Isopropanol	Merck	109634
KCl	Sigma-Aldrich	P9541
Laemmli buffer 4x	Biorad	1610747
Methanol	Merck	106009
MgCl$_2$	Sigma-Aldrich	M8266
Midori Green	Biozym	617004
Milk powder	Roth	T145
NaCl	Roth	9265.1
NaF	Roth	P756
NaOH	Applichem	A1551.1000
Nuclease-free water	Ambion	AM9937
PhosStop tablets	Roche	04906837001
Pluronic F-127	Thermo Fisher Scientific	P3000MP
Ponceau solution	Sigma-Aldrich	P7170
Protein gene ruler	Thermo Fisher Scientific	26619
Stainfree TGX gels	Biorad	4568086

SYBR green IQ supermix	Biorad	1708882
Tricine	Biochemika	A1085,0250
Tris base	Roth	5429.2
Triton-X-100	Sigma-Aldrich	3051.3
Turbo transfer (TB) buffer 5x	Biorad	10026938
Tween-20	Biorad	170-6531
β-Mercaptoethanol (cell culture)	Serva	28625
β-Mercaptoethanol (Western blot)	Biorad	1610710

2.2 COMMERICAL ANTIBODIES and KITS

Tab. 2 List of antibodies and kits.
IF: Immunofluorescence, WB: Western blot, FLOW: fluorescence cell cytometry, MACS: magnetic activated cell sorting

	Method	Company	Order number.
Primary Antibodies			
AFP	IF	Dako	A0008-4oC
CAMK2δ	WB/IF	Thermo Fisher Scientific	PA5-22168
CD31	IF/MACS	Thermo Fisher Scientific	MA1-26196
cTNT	IF/FLOW	Thermo Fisher Scientific	MS295PABX 13-11
GAPDH	WB	Millipore	MAB374
LIN28	IF	R&D systems	AF3757
MLC2v	IF	Proteintech	10906-1-AP
NANOG	IF	Abcam	PA5-18406
OCT4	IF	R&D systems	AF1759
PLN	WB	Thermo Fisher Scientific	MA3-922
PLN_Ser16p	WB	Badrilla	A010-12AP
PLN_Thr17p	WB	Badrilla	A010-13
RBM20	WB/IF	Myomedix	Order form
RYR2	WB	Sigma-Aldrich	HPA020028
RYR2_Ser2808p	WB	Badrilla	A010-30
RYR2_Ser2814p	WB	Badrilla	A010-31
SERCA2a	WB	Thermo Fisher Scientific	MA3-919
SOX2	IF	R&D systems	MAB2018
SSEA4	IF	Abcam	MC813
Titin M8/M9	IF	Myomedix	Order form
Titin Z1/Z2	IF	Myomedix	Order form
TRA1-60	IF	R&D systems	MAB4770
α-Actinin	IF	Sigma-Aldrich	A7811
α-SMA	IF	Sigma-Aldrich	A2547
β-III-TUB	IF	Covance	MMS-435P
Secondary Antibodies			
FITC goat-anti mouse IgM	IF	Jackson ImmunoResearch Laboratories	115-095-020

Cy3 goat-anti mouse IgG, IgM	IF	Jackson ImmunoResearch Laboratories	115-165-068
555 Donkey-anti goat IgG	IF	Thermo Fisher Scientific	A21432
555 Donkey-anti rabbit IgG	IF	Thermo Fisher Scientific	A31572
488 Donkey-anti mouse IgG	IF	Thermo Fisher Scientific	A21202
488 goat-anti mouse IgG, IgM	IF	Thermo Fisher Scientific	A10680
HRP Donkey-anti mouse IgG	WB	Thermo Fisher Scientific	A16011
HRP Donkey-anti rabbit IgG	WB	Thermo Fisher Scientific	A16023
Kits			
Click-iT EdU Cell Proliferation Kit	DNA/Proliferation	Thermo Fisher Scientific	C10637
Subcellular Protein Fractionation Kit	Cell fractionation	Thermo Fisher Scientific	78840
ALP staining Kit	ALP staining	Sigma-Aldrich	86R-1KT
Human stem cell nucleofector Kit 2	iPSC electroporation	Lonza	VPH-5022
NHDF Kit	Fibroblast electroporation	Lonza	VAPD-1001
Immobilon Western Kit	WB imaging	Millipore	WBKLS0500
QIAquick Gel Extraction Kit	Gel extraction	Qiagen	28706
SV total RNA isolation system	RNA Isolation of iPSCs	Promega	Z3105
Relia prep RNA tissue Miniorep Kit	RNA Isolation of iPSC-CMs	Promega	Z6112
QuantiNova RT Kit	Reverse transcription	Qiagen	205413
Pierce BCA Protein Assay Kit	Protein quantification	Thermo Fisher Scientific	23225
QIAamp DNA mini Kit	gDNA Isolation	Qiagen	51306

2.3 PRIMER

All primers were designed using the NCBI Primer Blast software and synthesized at Microsynth Seqlab. A complete list of primer sequences and mean Cq-values is provided in the supplements (Suppl. Tab. 1 and 2).

2.4 CELL CULTURE and CELL LINES

Tab. 3 List of media, supplements, factors and solutions for cell culture

	Company	Order number
Media and supplements for cell culture		
B27 supplement	Thermo Fisher Scientific	17504-044
DMEM without Glutamax	Thermo Fisher Scientific	11960-044
DMEM F12 with Glutamax	Thermo Fisher Scientific	31331028
Endothelial growth medium-2 (EGM2)	Lonza	CC-3162
Essential 8 (E8)	Thermo Fisher Scientific	A1517001
IMDM with GlutaMax	Thermo Fisher Scientific	31980022
RPMI 1640 with HEPES/GlutaMax	Thermo Fisher Scientific	72400021
RPMI 1640 without glucose	Thermo Fisher Scientific	11879020
StemFlex	Thermo Fisher Scientific	A3349401
StemPro-34 SFM	Thermo Fisher Scientific	10639011
Compounds and solutions for cell culture		
1-Thioglycerine (MTG)	Sigma-Aldrich	M6145-25ML
Albumin	Sigma-Aldrich	A9731
BMP4 human	R and D systems	314-BP-010
BSA Fraction 7.5 % solution	Thermo Fisher Scientific	152-037
CHIR99021	Merck	361559
EDTA 0.5 M pH=8.0	AppliChem	A3145.0500
EPO	Thermo Fisher Scientific	PHC2054
FCS	Thermo Fisher Scientific	10270-106
FLT3	Thermo Fisher Scientific	PHC9414
Geltrex	Thermo Fisher Scientific	A1413301
hbFGF	PeproTech	100-18B
HEPES solution	Sigma-Aldrich	H0887
Histofix 4 %	Roth	P087.5
IL-3	Thermo Fisher Scientific	PHC0034
IL-6	Thermo Fisher Scientific	PHC0065
Isoprenaline (Iso)	Sigma-Aldrich	I5627
IWP2	Merck	681671
Knock out serum	Thermo Fisher Scientific	10828-028
Lactate-solution 60%	Sigma-Aldrich	L9263
L-Ascorbic acid	Sigma-Aldrich	A8960
NEAA 100x	Thermo Fisher Scientific	11140035
PBS 10x	Thermo Fisher Scientific	70011-044
PBS 1x	Thermo Fisher Scientific	14190144
Penicillin/Streptomycin 100x	Thermo Fisher Scientific	15140122
SB431542	TORCIS	1614
SCF	Thermo Fisher Scientific	PHC2115
Thiazovivin	Merck	420220
Trypsin/EDTA 0.25	Thermo Fisher Scientific	25200056
VEGF-A$_{165}$	R and D systems	293-VE-010
Verapamil	Sigma-Aldrich	V4629
Versene (0.48 mM EDTA)	Thermo Fisher Scientific	15040066

Tab. 4 List of plasmids, viruses and CRISPR/Cas9 components for cell manipulation

	Company	Order number
Plasmids		
pCXLEh-Oct3/4-shp53-F	Addgene	27077
pCXLE-hSK	Addgene	27078
pCXLE-hUL	Addgene	27080
Viruses		
AAV6-GFP-U6-h-KHDRBS3-shRNA	Vector BioLabs	Upon request; Lot 180709#29
AAV6-GFP-U6-shRNA	Vector BioLabs	7043
Sendai virus	Thermo Fisher Scientific	A16517
CRISPR/Cas9		
Cas9 nuclease V3	IDT	1081058
crRNA	IDT	Individual design
Electroporation enhancer	IDT	1075916
HDR-template	IDT	Individual design
tracrRNA	IDT	1072532

Tab. 5 List of somatic cells and generated cell lines. * already established lines prior to project start. p. = reprogrammed with plasmids; s. = reprogrammed with Sendai virus

	Cell source	Lab intern ID
Somatic cells		
1-F *	Skin fibroblasts from LVNC patient 1 (LVNC pedigree II.3, p. 54)	6RBM
2-F	Skin fibroblasts from LVNC patient 2 (LVNC pedigree III.2, p. 54)	8RBM
3-PBMCs	PBMCs from DCM patient 1 (DCM pedigree III.6)	9RBM
4-F	Skin fibroblasts from DCM patient 2 (DCM pedigree III.7, p.54)	10RBM
5-F *	Skin fibroblasts from DCM healthy family member 1 (DCM pedigree II.2, p. 54)	3RBM
6-F *	Skin fibroblasts from DCM healthy family member 2 (DCM pedigree III.3, p. 54)	4RBM
1-C-F *	Skin fibroblasts from control 1	Ctl1
2-C-F *	Skin fibroblasts from control 2	Ctl2
3-C-PBMCs	PBMCs from control 3	Ctl3
4-C-PBMCs	PBMCs from control 4	Ctl4
5-C-F*	Skin fibroblasts	WTD2
iPSCs		
1-LVNC-1 *	iPSC-line 1 generated from fibroblasts 1-F of LVNC patient 1	p.6RBM1
1-LVNC-2 *	iPSC-line 2 generated from fibroblasts 1-F of LVNC patient 1	p.6RBM2
2-LVNC-1	iPSC-line 1 generated from fibroblasts 2-F of LVNC patient 2	s.8RBM1
2-LVNC-2	iPSC-line 2 generated from fibroblasts 2-F of LVNC patient 2	s.8RBM3
1-DCM-1	iPSC-line 1 generated from PBMC3-F of DCM patient 1	s.9RBM2
1-DCM-2	iPSC-line 2 generated from PBMC3-F of DCM patient 1	s.9RBM4
2-DCM-1	iPSC-line 1 generated from fibroblasts 4-F of DCM patient 2	s.10RBM3
2-DCM-2	iPSC-line 2 generated from fibroblasts 4-F of DCM patient 2	s.10RBM5
3-DCM-1 *	iPSC-line 1 generated from fibroblasts 5-F of DCM healthy 1	s.3RBM7
3-DCM-2 *	iPSC-line 2 generated from fibroblasts 5-F of DCM healthy 1	s.3RBM8
4-DCM-1 *	iPSC-line 1 generated from fibroblasts 6-F of DCM healthy 1	s.4RBM2
4-DCM-2 *	iPSC-line 2 generated from fibroblasts 6-F of DCM healthy 1	s.4RBM6
1-C-1 *	iPSC-line 1 generated from fibroblast of healthy donor 1	p.Ctrl1.1
2-C-2 *	iPSC-line 1 generated from fibroblast of healthy donor 2	p.Ctrl2.1
3-C-1	iPSC-line 1 generated from PBMC of healthy donor 3	s.Ctrl3.1
4-C-1	iPSC-line 2 generated from PBMC of healthy donor 4	s.Ctrl4.10
5-C-1 *	iPSC-line 1 generated from fibroblasts of healthy donor 5	p.WTD2
CRISPR/Cas9-edited iPSCs		
1 resLVNC 1	Isogenic line 1 generated from iPSC-line 1 LVNC 1	6cr-clone10
1 resLVNC 2	Isogenic line 2 generated from iPSC-line 1 LVNC 1	6cr-clone20
1 resLVNC 3	Isogenic line 3 generated from iPSC-line 1 LVNC 1	6cr-clone21
1 resLVNC 4	Isogenic line 4 generated from iPSC-line 1 LVNC 1	6cr-clone29
1 resDCM 1	Isogenic line 1 generated from iPSC-line 1 DCM 1	9cr-clone39
1 resDCM 2	Isogenic line 2 generated from iPSC-line 1 DCM 1	9cr-clone49

2.5 BUFFERS and SOLUTIONS

Tab. 6 Composition of buffers and solutions. Unless stated otherwise, the solvent was water.

Buffer or solution	Composition
1 % BSA/PBS	13.4 ml BSA solution (7.5 %) + 87 ml PBS (1x)
1.25 Tyrode solution	140 mM NaCl 4 mM KCl 1 mM $MgCl_2$ 10 mM HEPES 10 mM glucose 1.25 mM $CaCl_2$ (adjusted to pH = 7.4)
1.8 Tyrode solution	140 mM NaCl 5,4 mM KCl 1 mM $MgCl_2$ 10 mM HEPES 10 mM glucose 1.8 mM $CaCl_2$ (adjusted to pH = 7.4)
5 % milk and 5 % BSA	5 g milk or BSA powder + 100 ml TBST buffer
Anode buffer	36.4 g/l Tris base 17.9 g/l Tricine (adjusted to pH=8.8)
Caffeine solution (10 mM)	485.5 mg Caffeine + 125 ml water
Cathode buffer	36.3 g/l Amino caproid acid 3.6 g Tris base (adjusted to pH=8.7)
FACS buffer	1 % BSA 0.1 % Triton-X in PBS 1x
FACS-B10 buffer	0.5 % BSA in PBS 1x 10 % FCS in PBS 1x + 2.2 ml/l 0.5 M EDTA pH=8.0
Fluo-4 and FURA-2 solution	1 Tube (= 50 µg) Fluo-4 or FURA-2 + 44 µl DMSO
Iso stock solution (100 µM)	247.7 mg Iso + 10 ml water (= 100 mM) 1:1000 dilution of Iso (100 mM) in water
Laemmli buffer	900 µl Laemmli buffer 4x + 100 µl β-Mercaptoethanol
Running buffer	25 mM Tris, 192 mM Glycine, 0.1 % SDS
TBE buffer	10.8 g/l Tris base 5.5 g/l Boric acid 4 ml 0.5 M EDTA (pH=8) (for 1 l)
TBS buffer	20 mM Tris 150 mM NaCl (adjusted to pH=7.6)
TBST buffer	TBS buffer + 0.1 % Tween
Turbo transfer buffer (TB-buffer)	100 ml 5x Turbo transfer buffer + 400 ml water
Verapamil stock solution (100 µM)	491 mg Verapamil + 10 ml DMSO (= 100 mM) 1:1000 dilution of Verapamil (100 mM) in DMSO

3

METHODS

3.1 CELL CULTURE

IPSCs were generated from donated blood and skin samples. Subsequently, iPSCs were maintained in culture and utilized to differentiate into cardiomyocytes for experimental purposes. Tab. 7 summarizes the different media used in cell culture. The following chapters describe the process of reprogramming, cell culture maintenance and cardiac differentiation in more detail. All cell culture was maintained in an incubator at 37 °C, 5 % carbon dioxide and 95 % humidity.

Tab. 7 List of media used. * heat inactivated FCS (56 °C for 30 min).

Medium	Culture	Basis	Additives
Blood medium	PMBCs culture	StemPro	2 U/ml EPO 100 ng/ml FLT-3 100 ng/ml SCF 20 ng/ml IL-3 20 ng/ml IL-6
Human embryonic stem cell (hES) medium	iPSCs culture for EB formation	DMEM F12 with Glutamax	20 % knock out serum 1x NEAA 1x β-Mercaptoethanol
Human fibroblast medium (HFBM)	Fibroblast culture	DMEM without glutamax	10 % FCS * 1x L-Glutamine 1x NEAA 1x β-Mercaptoethanol 10 ng/ml hbFGF (B10)
Essential-8 (E8) Medium	iPSCs culture	Essential-8	E8 supplement (20ml/l)
StemFlex	iPSCs culture	StemFlex	StemFlex suppl. (20ml/l)
Cardio-culture medium	iPSC-CMs culture	RPMI with HEPES/Glutamax	B27 supplement with Insulin (20 ml/l)
Cardio-differentiation medium	iPSC-CMs differentiation start	RPMI with HEPES/Glutamax	500 mg/l Albumin 200 mg/l L-Ascorbic acid
Cardio-selection medium	iPSC-CMs metabolic selection	RPMI without HEPES and glucose	4 mM Lactate/HEPES 500 mg/l Albumin 200 mg/l L-Ascorbic acid

Medium	Culture	Basis	Additives
Cardio-digest medium	Digestion of iPSC-CMs	RPMI with HEPES/Glutamax	B27 supplement with Insulin (20 ml/l) 2 µM Thiazovivin 20 % FCS *
Iscove medium	iPSCs spontaneous differentiation	IMDM with Glutamax	20 % FCS * 1x NEAA 450 µM MTG
Freezing Medium	iPSCs culture	E8 Medium	20 % DMSO 4 µM Thiazovivin
Endothelial growth medium-2 (EGM2)	Endothelial cell culture	EGM2	EGM2 supplements (included with EGM2)

3.1.1 Isolation and culture of PBMCs from whole blood

Blood samples were sent to the laboratory in a standard EDTA- vacutainer for isolation of the PBMCs. Upon arrival, 5-10 ml of whole blood was transferred to a 50 ml tube and centrifuged 7 min at 600 g (without break). The yellow supernatant (plasma) was discarded and PBS (1:1) was added. The tube was inverted to mix, and the blood solution was carefully and slowly pipetted into a tube containing 5 - 8 ml Biocoll gradient (blood should not mix with the Biocoll). For separation of the blood cells, the Biocoll gradient was centrifuged 35 min at 400 g at 4 °C. The nebulous interphase containing the PBMCs (ca. 1 ml) was conferred to a 15 ml tube and washed once with PBS (centrifuged at 600 g for 5 min). The PBMC-pellet was resuspended in blood medium and transferred to one well of a 12-well plate for culture. The medium was refreshed daily by discarding half the of medium and replaced with fresh blood medium. After 5 - 7 d in culture, the PBMCs were used for reprogramming.

3.1.2 Isolation and culture of fibroblasts from a skin sample

Skin biopsies were sent to the laboratory in 15 mL tubes containing DMEM and 1x Penicillin/Streptomycin. The skin sample was cut with a scalpel into small pieces (1 – 2 mm^2), placed into a 6 cm dish with the subcutaneous layer facing down and the dermis facing up and cultured in human fibroblast medium (HFBM) (medium change every 2 – 3 d). After one – two weeks, fibroblasts growth out / around the skin sample was overserved. For passaging of fibroblasts, they were incubated with 0.25 % Trypsin/EDTA for 3 - 7 min at 37 °C. When the cells started to detach, the Trypsin/EDTA was carefully removed and HFBM was added to remove the fibroblasts from the culture dish by pipetting. The fibroblasts were transferred to a new 6 cm culture dish in a ratio of 1:6 – 1:14. Medium change with HFBM was carried out every 2 – 3 days. Isolated fibroblasts at passage 1 or 2 were used for reprogramming.

3.1.3 Reprogramming of fibroblasts and PBMCs

IPSCs were generated either from a blood or a skin sample donation. The reprogramming factors were transfected into the cells via Sendai virus (for fibroblasts and PMBCs) or via electroporation of episomal expressing plasmids (for fibroblasts).

For transfection with episomal plasmids, $0.5 - 1 \times 10^7$ fibroblasts (passage 1-2) were transferred to a 1.5 ml tube and centrifuged at 200 g for 10 min. The supernatant was discarded and the fibroblast pellet was carefully resuspended in 100 µl nucleofection solution containing 82 µl nucleofector solution and 18 µl nucleofector supplement (from NHDF kit). To these 100 µl, 1 µg of each plasmid pCXLEh-Oct3/4-shp53-F, pCXLE-hSK and pCXLE-hUL was added. Everything was blended gently and the mixture was transferred to a nucleofector cuvette (from NHDF kit) and gently tapped to remove air bubbles. The cuvette was placed into the Amaxa Nucleofector II and electroporation was carried out with the program P-22. The suspension from the cuvette was transferred to a 1/6 well with 1.5 ml HFBM containing 1x P/S and 2 µM Thiazovivin.

For transduction of fibroblasts with Sendai virus, fibroblasts with the same ratio were passaged into 2/12 wells. After 2 days, one of the wells was used to count the fibroblasts and assume the same cell number for the other well. With this cell number, the amount of virus was calculated to transduce the cells with an MOI of 10:10:6 (hKOS/hc-MYC/hKLF4). The calculated amount of Sendai virus was added to 1 ml HFBM with 1x Penicillin/Streptomycin and transferred to the well containing the fibroblasts and incubated for 24 h at 37 °C. The medium was changed with HFBM every second day. On day 7 post transduction, the fibroblasts were passaged onto a Geltrex - coated 6 well plate with the ratios: 1:4, 1:6, 1:8, 1:16, 1:20 and 1:24. The medium was changed to 1.5 ml E8 and replaced every 2 – 3 days. Colony growth was observed ca. 10 d post transduction.

For transduction with Sendai virus, after 5-7 days of cultivation 2×10^6 PMBCs were incubated in a 1.5 ml tube containing 300 µl blood medium and Sendai virus with an MOI of 4:4:4 or 10:10:6 (hKOS/hc-MYC/hKLF4). This was incubated for 3 h at room temperature and subsequently transferred onto a 1/12 well plate with 1 ml blood medium. After 1 day, the 1 ml was transferred to a Geltrex - coated 6 cm dish and 3 ml fresh blood medium was added. Every other day, half the medium was refreshed by carefully removing 2 ml from the surface. On day 7 post transduction, the medium was subsequently replaced with E8 medium and refreshed every 2 – 3 days. Formation of iPSC-colonies was observed 10-20 d post reprogramming and for every patient 6 colonies were picked and cultured as described in *2.1.5 Maintenance and culture of iPSCs*. 2 iPSC-clones were subsequently chosen at random for every patient.

The following formula was used to calculate the amount of Sendai virus used:

$$Virus\ volume\ [\mu l] = \frac{MOI\ [\frac{CIU}{cell}] * Number\ of\ cells}{Virus\ titer\ [\frac{CIU}{ml}] * 10\char94 - 3[\frac{\mu l}{ml}]}$$

The exact titer of the Sendai virus can be obtained from the analysis certificate (CIU = collective infectious units).

3.1.4 Coating of plates and dishes and cover slip preparation

For iPSCs and iPSC-CMs, the culture dishes were coated with Geltrex. For Geltrex - coating, 2 mg Geltrex were dissolved quickly in 12 ml cold DMEM F12 (167 mg/l) and 0.5 ml for 12 - well and 1 ml for 6 - well dishes were added and incubated for 30 – 60 min at 37°C. Plates with Geltrex were stored up to 1 week at 4°C. For experiments indicated, iPSCs or iPSC-CMs were digested onto round coverslips with a diameter of 20 mm for 12 - well and 25 mm for 6 - well plates. Before use, coverslips were incubated in 0.1 M hydrochloric acid for 1 h, for 30 min in 70 % Ethanol and subsequently cleaned with lint-free paper and sterilized for 2 h at 200 °C.

3.1.5 Maintenance and culture of iPSCs

IPSCs were cultivated on Geltrex - coated 6 - Well dishes containing 1.5 ml E8 medium that was changed daily. When iPSC reached 80 - 90 % confluency, they were passaged in a ratio of 1:6 – 1:12 (depending on the cell line) to a dish with 1.5 ml E8 medium containing 2 µM Thiazovivin, a ROCK inhibitor. For this, the iPSCs were detached by adding 1ml Versene for 5 -7 min at 37 °C. When cells started to detach, the Versene was carefully removed and the iPSCs were resuspended in 1 ml E8 / Thiazovivin medium and transferred to a new Geltrex - coated well. IPSCs are sensitive to oxidative and mechanical stress, therefore only 2 - 3 passes for resuspension were pipetted.

3.1.6 Freezing and thawing of iPSCs

To freeze iPSCs, the cells were treated with 1 ml Versene for 5 -7 min at 37 °C. Versene was removed and the iPSCs carefully resuspended in 500 µl E8 medium. 500 µl freezing medium was carefully added to the cells and subsequently transferred to a cryotube and slowly frozen in Mr. Frosty (Nalgene) containers at -80 °C. Afterwards, the cryotubes were placed in liquid nitrogen tanks for long-term storage.

To thaw iPSC, the frozen cryotubes were rapidly thawed in a 37 °C water bath and the cells pipetted into a 15 ml tube containing 5 ml E8 medium. The cells were then centrifuged at 200 g for 5 min and the supernatant was discarded. The cell pellet was mixed in 1.5 ml E8 with 2µM Thiazovivin and transferred to a Geltrex - coated 1/6 well.

3.1.7 Spontaneous differentiation of iPSCs

To verify pluripotency of the iPSC-lines, they were cultured in Embryoid bodies (EBs) to induce spontaneous differentiation of the three germlayers. For this, the iPSCs were passaged onto Mitomycin C inactivated mouse embryonic feeder (MEF) layer cells in a 6 cm dish with hES medium with hbFGF (10 ng/ml). When the cells reached 80 – 90 % confluency, the cells were washed twice with DMEM F12 medium and incubated with collagenase IV (200 U/ml) for 5 min at 37 °C for detachment. The cells were mechanically detached into cell clumps using a cell scraper. These cell clumps were transferred onto an uncoated 6 cm dish with hES medium (d0 of spontaneous differentiation). The following day (d1), the medium was replaced with Iscove medium to induce differentiation and subsequently changed every other day. At d8, 16-18 EBs were collected as a pellet for PCR analysis. The remaining EBs were transferred to Geltrex - coated 12 well with and without coverslips (5-8 EBs per 12 well). On day 8+8, 2 wells with coverslips were fixated for AFP staining. On day 8+25, 5-6 wells with coverslips were fixed for α-SMA and β-III-TUB staining. In parallel, the EBs in the 12 well plate without a coverslip were collected as a pellet for PCR analysis.

3.1.8 Directed differentiation into endothelial cells (ECs)

For directed differentiation into ECs, a previously published protocol was used (Natividad-Diaz et al., 2019). For this, the iPSCs were plated onto Geltrex - coated 12 - well dishes. When they reached a confluency of 80 - 90 %, the differentiation was started by replacing the medium with 2 ml E8 medium containing 6 µM CHIR99021 (d0).). After 2 days (d2), the medium was changed to E8 medium with BMP4 (10 ng/ml), VEGF-A (50 ng/ml) and SB431542 (10 µM). This medium was refreshed at d4. At d5, the medium was replaced with EGM2 medium. On d7, the iPSC - ECs were magnetically sorted with CD31 antibodies. For this, the iPSC-ECs were detached with 0.25 % Trypsin/EDTA for 5 min and counted. Dynabeads, in a ratio of 5:1 beads to cells, were washed and resuspended in 200 µl FACS-B10 buffer and incubated with 4 µl CD31 antibody (1:50) for 30 min at RT. Subsequently, the CD31-dynabeads were washed twice with FACS - B10 buffer and resuspended in 500 µl. The detached iPSC - ECs were centrifuged at 200 g for 5 min and the cell pellet was resuspended in 500 µl CD31 - Dynabeads and incubated at RT for 30 min. Subsequently, the CD31+ cells bound with CD31 -Dynabeads were washed twice with FACS - B10 buffer and once with EGM2 medium. The bead - bound cells were transferred into a Geltrex - coated 1/6 well plate containing 2 ml EGM2 medium with 2 µM Thiazovivin. The medium was replaced with EGM2 the following day and refreshed every other day. When the cells reached 90 % confluency, they were passaged in a ratio of 1:12.

3.1.9 Directed differentiation into ventricular iPSC-CMs

For directed cardiac differentiation into ventricular iPSC-CMs, a previously published protocol was adapted (Lian et al., 2013; Tohyama et al., 2013). For this, the iPSCs were plated into Geltrex - coated 12 -well dishes. When they reached a confluency of 80 - 90 %, the differentiation was started by replacing the medium with 2 ml cardio-differentiation medium containing 4 µM CHIR99021 (d0). After 2 days (d2), the medium was changed to cardio-differentiation medium containing 5 µM of IWP2, an inhibitor of the Wnt-pathway. On d4 and d6, the medium was changed with 2 ml cardio-differentiation medium and switched to cardio-culture medium on d7. Beating cells were observed between d8 – d10. At d14 – d20, iPSC-CMs were detached from the 12 - well plate using 1 ml 0.25 % Trypsin/EDTA for 5 min and counted with a Neubauer chamber. 6×10^6 iPSC - CMs were plated in Geltrex-coated 6 - well dishes in 2 ml cardio-digest medium. After 2 days, the medium was replaced with a cardio - selection medium to metabolically select for iPSC-CMs. This selection step was carried out for 5 days with 2 ml cardio-selection medium change every second day. Afterwards (d25 – d28), the medium was switched back to 2 ml of cardio-culture medium, which was replaced every 3 – 4 days. The iPSC-CMs were cultivated for 60 – 90 days before being used in experiments unless stated otherwise.

3.1.10 Directed differentiation into atrial-CMs

Differentiation into atrial iPSC-CMs was performed by Wiebke Maurer. For directed cardiac differentiation into atrial iPSC-CMs, a previously published protocol was used (Cyganek et al., 2018). The protocol is identical to the protocol for ventricular iPSC-CMs differentiation (see chapter 3.1.9 *Directed differentiation into ventricular iPSC-CMs*), except for the addition of 1 µM retinoic acid (RA) at day 3 (IWP2 + RA) and day 4-6 (RA only).

3.1.11 Digestion of iPSC-CMs

For different experiments, the iPSC-CMs were digested onto different wells or onto coverslips with different densities depending on the experiment. Densities and cell number are summarized in Tab. 8. For digestion, the iPSC-CMs are incubated with 1 ml 0.25 Trypsin/EDTA at 37 °C. After 5 – 8 min, 2 ml cardio - digest was added and the cells were detached by pipetting and transferred to a 15 ml tube. The cells were centrifuged for 5 min at 200 g and the pellet resuspended in cardio - digest medium. At this step, the iPSC-CMs were counted if an exact cell number was necessary. The designated amounts of iPSC - CMs were pipetted into Geltrex - coated culture plates or dishes and 2 ml cardio - digest medium was added. 2 days later, the medium was changed to cardio-culture and refreshed every 2 – 3 days.

Tab. 8 Numbers of iPSC-CMs used. CS: cover slip

Experiment	Dish/plate	Cell number	confluency
IF of iPSCs	12 well with 20 mm CS	Passage: 1:10	Single iPSC colonies
Sarcomere analysis and IF of iPSC-CMs	12 well with 20 mm CS	$3 - 4 \times 10^4$	Single cells/ small groups
Ca^{2+} with Fluo-4	6 well with 25 mm CS	$2.5 - 3 \times 10^5$	Confluent monolayer
Ca^{2+} with FURA-2	35 mm Fluoro dish	$3 - 4 \times 10^4$	Single cells
Multi-electrode assay (MEA)	6 well MEA plate	3×10^5	Confluent/monolayer
EdU-Assay	12 well with 20 mm CS	1×10^5	Confluent/monolayer

3.1.12 Pellet collection and fixation of cells

To collect cells for further experiments, they were either taken as a cell pellet or fixed on coverslips. For cell pellet collection, iPSCs, iPSC-CMs or iPSC-ECs were washed once with PBS and subsequently mechanically scraped in 1 ml PBS and transferred into a 2 ml tube. The tube was centrifuged 1 min at 13.000 g to form a cell pellet. The PBS supernatant was discarded and the tube was immediately shock frozen in liquid nitrogen. Frozen cell pellets were stored at -80 °C.

To fix iPSCs or iPSC-CMs on coverslips, the cells were washed once in PBS and then fixed with Histofix. For this, the cells were incubated with 1 ml Histofix for 20 min at RT. Afterwards, the Histofix was discarded into a special Formalin waste bin and the cells were washed once with PBS. Afterwards 1.5 ml 1 % BSA/PBS was added and the plates were stored at 4 °C until further use.

3.1.13 AAV6 mediated SLM2 knock down

An AAV6 mediated approach was chosen to knock down SLM2 in control iPSC-CMs. An MOI of 10^5 was used for the AAV6-shRNA-scrambeld and AAV6-SLM2-shRNA vector. 5×10^4 control iPSC-CMs were passaged onto Geltrex - coated 20 mm round coverslips. The cells were incubated with the respective AAV6 for 3 days. Subsequently, the medium was changed twice a week with cardio - culture medium supplemented with 1x Penicillin/Streptomycin. At d14, after transduction start, the cells were fixed and stained for the sarcomeric Z-disc with the titin-Z1/Z2 antibody (1:500).

3.2 MOLECULAR BIOLOGY

3.2.1 Genomic-DNA and RNA Isolation

Genomic DNA (gDNA) was isolated from cell pellets using the QIAamp DNA Kit following the manufacturer's "protocol for cultured cells". For RNA isolation of iPSCs, the SV total RNA-Isolation Kit and for RNA isolation of iPSC-CMs the Promega Relia Prep RNA tissue System Kit was used. Following the manufacturer's instructions, a 30 min incubation step with DNAse I for gDNA digestion during RNA isolation was included and was carried out when samples were used for next generation sequencing (NGS). RNA isolation from cells being subjected to PCR assays excluded the DNAse I and DNAse Stop buffer step during the isolation protocol since an efficient gDNAse digest was included in the reverse-transcriptase protocol used (see next paragraph).

3.2.2 Reverse transcription of RNA

For analysis in PCR, the isolated RNA was transcribed into cDNA by reverse transcription (RT) using the enzyme reverse transcriptase. The QuantiNova RT Kit provided gDNase, RT-Enzyme and an RT-buffer. For this purpose, 100 – 300 ng RNA in a volume of 13 µl (supplemented with water) were incubated with 2 µl gDNA Removal Mix for 2 min at 45 °C. Immediately after the 2 min incubation step, 5 µl RT-Mix containing 4 µl RT-buffer and 1 µl RT-Enzyme (1 U) were added and incubated with the following settings: 3 min at 25 °C → 10 min at 45 °C → 5 min at 85 °C. Depending on the amount of RNA used in the RT, water was added after the RT process to obtain constant RNA concentrations: 100 ng (no water added), 200 ng (20 µl water added) and 300 ng (40 µl water added).

3.2.3 Semi-quantitative and quantitative PCR

For semi-quantitative PCR, 10 ng/2 µl cDNA were mixed with 5 µl Green buffer (5x), 1 µl forward and reverse primer (10 µM), 1.6 µl dNTP mix (10 mM), 0.1 µl Tag-polymerase and 14.3 µl water (end volume 25 µl). The PCR mix was subjected to the following cycling conditions:

1. 5 min 95 °C   (initial denaturation)
2. *30 s 95 °C*   *(Denaturation)*
3. *20 s 55-60 °C* *(Primer annealing)*
4. *30-60 s 72 °C* *(Primer elongation)*
5. 10 min 72 °C   (Final elongation)
4 °C   (cooling until use)

Step 2-4 was repeated 40 times. Annealing temperature depends on the respective primer pair and the elongation time on the expected PCR product (ca. 10s per 100 bp). The PCR product was stored at 4 °C until it was used for gel electrophoresis.

Quantitative PCR (qPCR) was performed using the Biorad CFX Connect system. The qPCR reaction mix contained 5 ng/ 1µl cDNA, 10 µl SYBR green mix (2x,), 1 µl of forward and reverse primer (10 µM) and 8 µl water to yield a reaction volume of 20 µl. The qPCR mix was pipetted into a well of a 96 - well plate as duplicates for every sample. The following cycling conditions were used:

1. 3 min 95 °C   (initial denaturation)
2. *20 s 95 °C*   *(denaturation)*
3. *20 s 60 °C*   *(primer annealing)*
4. *30 s 72 °C*   *(Primer elongation)*
5. 10 s 95 °C
6. 60 °C to 95 °C in 0.05 °C steps (melt curve)

Step 2 - 4 was repeated 40 times. The Cq-value, which is the first cycle detected above the background threshold, is calculated by the machine. To calculate relative amounts of mRNA the delta-ct method was applied. For relative gene amounts the housekeeping gene *18s* or *GAPDH* was used. In the splice assays, the respective gene amounts (primer spanning a constitutive exon) was used for normalization. To monitor the PCR, 2 negative controls were utilized, which comprised a water sample (to test PCR components) and a "-RT" sample (to test gDNA background). A result of Cq > 35 was considered negative.

3.2.4 Agarose gel electrophoresis and gel extraction

Agarose gel was prepared depending on the expected length of the PCR-fragments: 100 – 500 bp (1.5 %) and >500 bp (2.0 %) containing 6 µl/100 ml Midori Green. The agarose powder was mixed in TBE-buffer and boiled 3 times to dissolve completely. After cooling at RT for 10 min, Midori Green (0.07 µl/ml) was added. 20 – 15 µl from PCR reaction were loaded in one pocket and 10 µl 100 bp Gene ruler was loaded to monitor PCR product size. The PCR fragments were separated with 100 – 120 V for 30 – 60 min and visualized under UV light. If the PCR products were subsequently subjected to Sanger sequencing (Microsynth Seqlab), the cDNA bands were cut preciously from the gel and purified using the QIAquick Gel Extraction Kit following the manufacturer's instructions, including the additional washing steps recommended for subsequent sequencing applications. To increase the DNA yield, the elution volume was decreased to 20 – 30 µl water.

3.2.5 Gene editing with CRISPR/Cas9 technology

To edit the RBM20 mutation in LVNC (c.G1901T) and DCM (c.C1900T) back into wt-RBM20, HDR-based CRISPR/Cas9 technology was applied. The crRNAs were designed to target the Cas9 to the RBM20 locus and introduce the DSB closest to the mutation using the Custom Alt-R© CRISPR-Cas9 guide RNA design tool from IDT (https://eu.idtdna.com/site/order/designtool/index/CRISPR_CUSTOM). The following crRNAs were chosen: 5′-CTCACCGGACTACGAGACAG-3′ for LVNC and 5′-GATATGGCCCAGAAAGGCCG-3′ for DCM. The HDR template was designed to harbor wt-RBM20 at position p.634/c.1900-1902 and artificially introduce silent mutations at the PAM and crRNA binding site to enhance editing efficiency. The HDR-template is a ssDNA with 5′-GGTGTGAAGATTCTAAATCCTGCTCCTTGGCTCCCTCACAGATATGGCCCAGAAAGGC CGCGGTCTCGTAGTCCCGTCAGCCGGTCACTCTCCCCGAGGTCCCACACTCCCAGCT TCACCTCC-3′ for LVNC and 5′-GGTGTGAAGATTCTAAATCCTGCTCCTTGGCTCCCTCACAGATATGGACCAGAAACGGT CTCGTAGTCCGGGAGCCGGTCACTCTCCCCGAGGTCCCACACTCCCAGCTTCACCT CC-3′ for DCM. The CRISPR strategy is visualized in Fig. 7A-C. The crRNA, Cas9 protein and ssDNA-HDR-template were directly electroporated into the iPSCs. For this, the following steps were performed:

1. Assembly of crRNA and tacrRNA: 5 µl cRNA (100 µM) with 5 µl tacrRNA (100 µM) were mixed and incubated at 95 °C for 5 min. Subsequently, the mix was cooled at RT, which allows to crRNA-tacrRNA-complex to assemble.

2. Assembly of Cas9 protein with the crRNA-tacrRNA-complex: 2 µl Cas9 protein (10µg/ml) were mixed with 6 µl crRNA-tacrRNA-complex from (1.) and incubated for 20 min at RT.

3. Preparation of electroporation mix: 8 µl of crRNA-tacrRNA-Cas9 from (2.) were mixed with 82 µl nucleofector solution, 18 µl nucleofector supplement (from Human Stem cell Nucleofector Kit 2), 1 µl electroporation enhancer, 3 µl HDR-template (100 µM) and incubated for 5 min at RT. Subsequently, the mix was heated to 37 °C in preparation of electroporation.

4. iPSCs were pretreated with E8 medium with 2 µM Thiazovivin for 1 h. The cells were detached using Versene and counted. $2x10^6$ iPSCs were centrifuged at 200 g for 5 min. The pellet was dissolved in the electroporation mix from (3.) and transferred into nucleofection cuvette (from the Human stem cell nucleofector Kit 2).

5. Electroporation: The nucleofection cuvette was placed into the Amaxa Nucleofector II and the program B - 016 was used for electroporation settings. Afterwards, the solution in the cuvette was carefully pipetted (excluding the foam) into Geltrex - coated 4 / 6 well plate containing StemFlex with 2 µM Thiazovivin.

6. Singularization: After 2 d, the cells were singularized by digesting a small cell number into a new Geltrex - coated 6 well plate with densities of: 800, 1000, 1200, 1500, 1800 and 2000 cells. Medium change with StemFlex was performed every other day.

Growth of iPSC - colonies was observed 7 – 10 d later and 72 colonies were picked and transferred into a 1 / 6 well plate. When the well reached 80 – 90 % confluency, half the well was frozen and the other half collected as a pellet for gDNA isolation and Sanger sequencing.

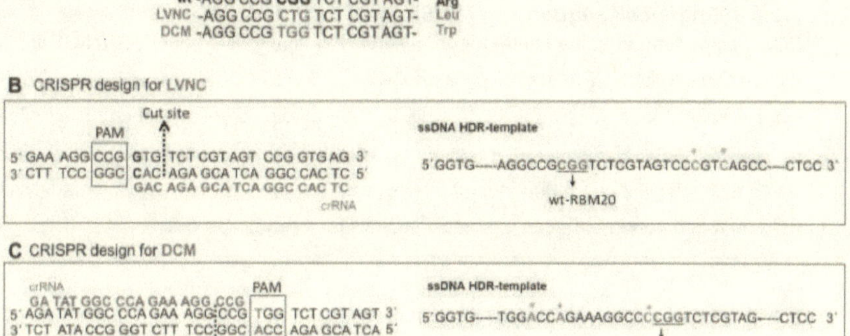

Fig. 7 CRISPR design for RBM20 gene editing in LVNC and DCM-iPSCs
A: The RBM20 locus of wt, LVNC and DCM with the respective nucleotide exchange for the p.R634 position.
B: CRISPR design for the LVNC. * Silent mutations in the HDR template
C: CRISPR design for the DCM. * Silent mutations in the HDR template

3.2.6 Sequencing

3.2.6.1 Sanger Sequencing

Sanger sequencing was performed at Microsynth Seqlab, Göttingen. For screening of RBM20 edited iPSC - clones, gDNA was isolated and the RBM20 locus amplified by PCR. The PCR reaction was sent to Microsynth Seqlab, where PCR purification was performed prior to sequencing. For identification of splice isoforms, RT-PCR was performed and the PCR bands excised and purified from the agarose gel. 10 – 40 ng/µl DNA (25 µl) were sent with the respective primer (10 µM) to Microsynth Seqlab.

3.2.6.2 Next-generation sequencing (NGS)

NGS was performed in cooperation with the AG Meder, Heidelberg. For this, mRNA of 90d old iPSC-CMs from control, LVNC-, DCM-, resLVNC- and resDCM-CMs were isolated. For every patient, 3 differentiation experiments were used. The mRNA was sent to Heidelberg, where NGS and analysis of the fastq files was performed by Dr. Weng-Tein from the AG Meder (Fig. 8). The analyzed files were sent back to Göttingen and used in GOterm analysis to identify novel splice targets (see chapter "*3.2.7 Identification of novel splice targets*").

3.2.6.3 Single-cell sequencing (SCS)

SCS was performed at the center for NGS-service for integrative genomics (NIG), Göttingen (Dr. Gabriela Salinas). For this, 30 d old iPSC-CMs of LVNC-, DCM, resLVNC- and DCM-CMs were prepared in a volume of $1x10^6$ cells in 1 ml cardio - digest medium. The cells were transported to the NIG, where they were singularized and sequenced (paired end reads) by iCell8 technology. A list with genes of interest was provided to the NIG, which analyzed the SCS data and generated gene expression results for every single cell.

3.2.7 Identification of novel splice targets

To identify novel RBM20 splice targets, 36 candidate genes were selected from the NGS data. The analysis of the fastq files was performed by the AG Meder as described above. A list with differential spliced exonic bins between control/resLVNC vs LVNC and control/resDCM vs DCM was provided. The following steps were applied: 1.) To minimize possible targets, all exonic bins with $p_{adjusted}$ value > 0.05 were discarded. 2.) All genes with differential exonic bin expression of control/resLVNC vs LVNC and control/resDCM vs DCM were listed using Venny2.1 software. This identified unique and shared genes with differential exon usage. 3.) The list of shared genes (RBM20 will affect the same gene but different exons) was extracted and subjected to GOterm analysis with the term "Human Diseases" (pvalue = 0.05). Genes with cardiac function were selected from the GOterm "Cardiomyopathy". 5.) From the selected genes, the exonic bins with $p_{adjusted}$ > 0.05 and fold change <+ /- 1 were selected and translated to the respective exons. 6.) Primers were designed to span the exon of interest and screened by PCR and gel electrophoresis to assess if multiple PCR products can be detected. If multiple bands were detected, they were subjected to Sanger sequencing.

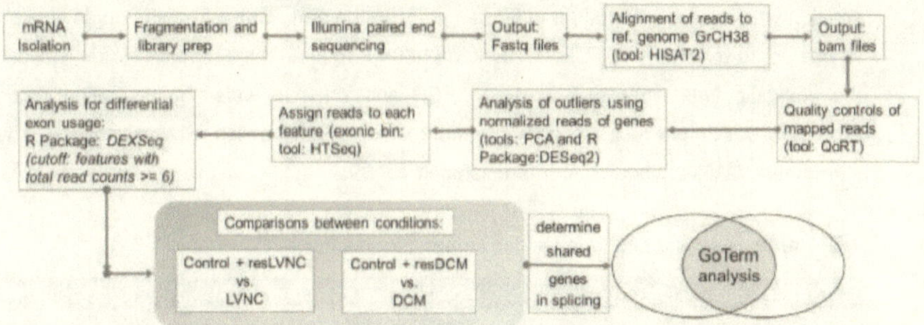

Fig. 8 Analysis pipeline for NGS data to detect novel RBM20 splice targets.

3.2.8 Alkaline phosphatase (ALP) staining

Activity of ALP is a general marker for stem cells. To test if the iPSC-lines have ALP activity the Alkaline Phosphatase Kit was used. When the iPSCs reached 30 - 40 % confluency they were washed once with PBS and covered with 1 ml fixation solution (5.8 ml citrate solution (provided by the Kit), 13 ml acetone and 1.2 ml 37 % formaldehyde) for 30 s. The fixation solution is discarded into the Formalin waste bin and the cells washed with water twice. The ALP staining solution was prepared as follows: 1 volume of FRV (from Kit) was mixed with 1 volume of sodium nitrate (from Kit). After 1 min at RT, 45 volumes of water and 1 volume of Napthol (from Kit) was added and mixed. The cells were incubated with 1 ml staining solution at 37 °C for 10-20 min. The staining solution was discarded, the cells washed twice with water and left to air dry. Stained cells were stored at RT until imaged.

3.2.9 Immunofluorescence (IF)

IF can visualize proteins in a cell. For this, specific antibodies are used to bind the protein/antigene of interest and a second fluorochrome - labelled and species - specific antibody is used to visualize the first bound antibody. For IF-staining, iPSCs or iPSC-CMs were digested onto Geltrex - coated 20 mm round coverslips placed in a 12 - Well plate. IPSCs were fixated when desired confluency was reached. For iPSC-CMs $3-4 \times 10^4$ cells were digested onto the coverslip and fixated after for 7 – 9 days. Afterwards cells were washed once in PBS and 2 ml 1 % BSA/PBS was added to each well and stored at 4 °C until use. Blocking step was carried out with 1 % BSA/PBS for 2 h at RT or ON at 4 °C. The coverslip was removed from the 12 - well, washed in a PBS filled weighing pan and placed on a cap in a wet chamber. 200 µl of the primary antibody solution were added on top and incubated ON at 4 °C. The coverslip was then washed 3 times in PBS and 200 µl of the secondary antibody solution was added and incubated 60 min at RT in the dark. Subsequently the coverslip was incubated 10 min with HOECHST solution (1:5000) to stain the cell nuclei. After incubation, the coverslip was washed 3 times in PBS and once in water before being mounted upside - down onto a microscope slide and sealed with nail polish. Stained coverslips were stored at 4 °C. All antibodies fir IF are listed in Tab. 9.

Tab. 9 Antibodies and dilutions used in Immunofluorescence.

Primary Ab	Produced	Dilution	Secondary Ab	Dilution
Staining of iPSCs				
OCT4	goat, IgG	1:40	555 donkey – anti goat	1:1000
SOX2	Mou, IgG	1:200	Cy3 goat - anti mouse	1:300
NANOG	Rb, IgG	1:100	555 donkey - anti rabbit	1:1000
LIN28	goat, IgG	1:300	555 donkey - anti goat	1:1000
TRA1-60	Mou-IgM	1:200	488 donkey – anti mouse	1:200
SSEA4	Mou, IgG	1:200	488 donkey – anti mouse	1:1000
Staining of EBs				
AFP	Rb, IgG	1:500	555 donkey - anti rabbit	1:1000
β-III-Tubulin	Mou, IgG	1:2000	488 donkey – anti mouse	1:1000
α-SMA	Mou, IgG	1:3000	488 donkey – anti mouse	1:1000
Staining of iPSC-CMs				
cTNT	Mou, IgG	1:500	488 donkey – anti mouse	1:500 (Flow Cytometry)
Titin-Z	Rb, IgG	1:750	555 donkey - anti rabbit	1:750
Titin-M8/M9	Rb, IgG	1:750	555 donkey - anti rabbit	1:750
a-Actinin	Mou, IgG	1:1000	488 donkey – anti mouse	1:1000
MLC2v	Rb, IgG	1:200	555 donkey - anti rabbit	1:200

3.2.10 Western blot

3.2.10.1 Protein Isolation from frozen iPSC-CMs pellets

The cell pellet was resuspended on ice in 80 – 100 µl protein lysis buffer containing 20 mM Tris-HCl (pH = 7.4), 200 mM NaCl, 20 mM NaF, 1 % Igepal, mM Na_3VO_4, 1 mM DTT, water and 1x Phosphostopp and 1x protease inhibitor. The pellet was resuspended by pipetting, followed by a 30 min incubation on ice and subsequently centrifuged at 4°C at 3000 g for 8 min. The supernatant was carefully transferred to a new 1.5 ml tube and stored at -80 °C. To assess the protein concentration, the protein lysate was diluted 1:20 in PBS. Next, 3 times 25 µl of each sample was mixed with 196 µl Solution A and 4 µl Solution B from the Pierce™ BCA Protein Assay kit in a 96-well format and incubated 30 min at 37°C. The protein concentration was measured at 562 nm emission. The protein samples were stored at -80°C until use.

3.2.10.2 SDS-polyacrylamide electrophoresis, protein transfer and detection

To separate the protein samples by size, they were denaturized and subsequently separated by electrophoresis. Protein [1.35 µg/µl] in PBS was mixed with Laemmli buffer and incubated for 30 min at 37 °C to denature the protein. 15 µl of sample were loaded in each lane of a pre-cast MINI-PROTEAN TGX stain-free gel. The electrophoreses chamber was filled with running buffer. Depending on the number of gels in the electrophoresis chamber, constant 15-20 mA/gel was applied. The electrophoresis was terminated when the dye front reached the end of the SDS-gel. The SDS-gels were carefully removed and placed into the Biorad Chemidoc XRS imager for photoactivation of the fluophores (setting "Stain free Gel"). The stain free technology allows monitoring protein amounts on the gel and after transfer, because during the gel-electrophoresis Trp residues are covalently bound to a proprietary trihalo moiety, which gives a fluorescent signal under UV light. Afterwards, a semi - dry transfer was prepared using either a nitrocellulose membrane (pore size 0.45 µm) in combination with TB-buffer or methanol-activated PVDF membrane (pore size 0.45 µm) with anode and cathode buffer (see Tab. 10). In both cases, the Whatmann paper, membrane and gel were equilibrated in the respective buffer before the transfer chambers were assembled. For transfer, the Trans – Blot® Turbo™ system was used, which allows a fast protein transfer from the gel onto the membrane. Since the transfer time depends on protein size, the membrane and gels were cut to adjust different transfer times according to the protein size (see Tab. 10) and were run at 25 V. Afterwards, the membranes were washed in water and TBST buffer and imaged with the setting "Stain free Blot" to detect whole protein content on the membranes after the transfer. The membranes were blocked for 1 h at RT in 5 % BSA or 5 % milk and subsequently incubated with the primary antibody diluted in 1 % BSA or 1 % milk at 4°C ON. After incubation, the membrane was washed 3x times in TBST buffer followed by incubation with the respective secondary antibody (coupled with horse reddish peroxidase) for 1 h at RT. The membrane was washed again 3x times in TBST buffer and imaged with the setting "colorimetric" to capture the protein marker. Afterwards, the membrane was incubated with HPR substrate for 5 min at RT to visualize the protein bands. The setting "Chemi high sensitivity" was chosen with "Signal accumulation mode" to capture an image every 5 s for 5 min. The best image was selected and the band intensity was quantified using Image Lab. Afterwards, the membranes were washed in TBST buffer and stripped with Ponceau red solution for 1 h. Subsequently, the membranes were incubated with another antibody or stored at -20 °C.

Tab. 10 Antibodies and dilutions used in Western blot. Rb = rabbit, Mou = mouse

Antibody		Dilution	Size	Transfer: Membrane/time
CAMK2δ	Rb, IgG	1:5000 BSA	55 kDA	PVDF/3-4 min
CAV1.2	Rb, IgG	1:250 BSA	160/110/80 kDa	Nitrocellulose/ 14 min
GAPDH	Mou, IgG	1:500 milk	38 kDa	PVDF/3 min
PLN	Mou, IgG	1:5000 milk	5-10 kDa	PVDF/3 min
PLN_Ser16p	Rb, IgG	1:5000 milk	5-10 kDa	PVDF/3 min
PLN_Thr17p	Rb, IgG	1:5000 milk	5-10 kDa	PVDF/3 min
RBM20	Rb, IgG	1:750 BSA	180 kDa	Nitrocellulose/ 14 min
RYR2_Ser2808p	Rb, IgG	1:1000 BSA	560 kDa	Nitrocellulose/ 14 min
RYR2_Ser2814p	Rb, IgG	1:4000 BSA	560 kDa	Nitrocellulose/ 14 min
RYR2	Rb, IgG	1:5000 milk	560 kDa	Nitrocellulose/ 14 min
SERCA2a	Mou, IgG	1:20000 milk	110 kDa	Nitrocellulose/ 14 min

3.2.11 Flow cytometry (FLOW)

To quantify the differentiation efficiency, the percentage of CMs was quantified by FLOW analysis using the cardiac marker cTNT. For this, 0.3-1x10^6 cells were detached with 0.25 trypsin / EDTA treatment for 10 min. The cells were centrifuged 5 min at 200 g and washed once with PBS. Subsequently, the cells were fixed with Histofix for 20 min at RT and afterwards washed twice with PBS. At this step the cells were split into 3 tubes to prepare a blank sample, secondary antibody control and a cTNT sample. The blank and secondary antibody sample were stored in FACS buffer until further use. For the cTNT sample, the cells were incubated ON at 4 °C with the primary antibody cTNT in a 1:500 dilution in FACS buffer (1 % BSA / PBS with 0.1 % Triton-X). The following day, the cTNT-cells and the secondary antibody sample were washed 3 times with FACS buffer and incubated 1 h at RT with the secondary antibody 488 donkey-anti mouse (1:500). Subsequently, all 3 tubes (blank, secondary control and cTNT sample) were washed 3 times with 0.2 % BSA/PBS and resuspended in 500 µl PBS. The cells were measured with the FACS Canto II with the following settings: 10000 events, forward scatter 228 V, side scatter 440 V, 488 Alexa Fluor laser 390 V. The detection threshold was set according to the blank sample and cTNT positive cells were presented as cTNT positive cells in [%].

3.3 FUNCTIONAL ASSAYs

3.3.1 Multi-electrode assay (MEA)

The MEA allows digesting beating iPSC-CMs onto electrodes to measure their electrical activity. The signal comprises a Na^+ spike and t-wave signal from which the beating rate (BR), field potential duration (FPD) and interspike interval (ISI) can be derived. The FPD was beat rate corrected using the Fridericia correction (cFDP) (Vandenberk et al., 2016). Custom made 6 - well MEA plates were stored in water at RT. On the day of usage, the 6 - well MEA was sterilized using a 70 °C water bath for 30 min and UV incubation for additional 30 min. Following the sterilization, a 7 µl Geltrex drop was placed directly onto the electrodes in the center of the MEA and incubated 30 min at 37 °C. The iPSC-CMs were trypsinized and the pellet dissolved in a small volume (0.7 – 1 ml) of cardio - digest medium. The Geltrex was removed and a drop of cardio - digest medium containing $2.5x10^4$ CMs was carefully pipetted onto the electrodes and incubated for 20 min at 37 °C to allow the cells to attach onto the electrode region. Afterwards, 350 µl cardio digest medium was added in every well and changed every 2 - 3 d with cardio - culture. The iPSC-CMs were cultured at least 7 d before MEA recording. To record and analyze the MEA signals, the MCRack software was used. The following conditions were recorded: Basal beating for 5 min, stimulated beating with 100 nM Iso was recorded for 10 min (after 5 min incubation). Medium was changed and after 24 h and basal beating recovery was recorded for 5 min.

3.3.2 Ca^{2+} imaging

To analyze Ca^{2+} handling, the cells were stained with Fluo-4, a cytosolic Ca^{2+} dye to analyze Ca^{2+} kinetics and sparks. To complement the Ca^{2+} analysis, the iPSC-CMs were stained with FURA-2, a ratiometric cytosolic Ca^{2+} dye that allows to analyze Ca^{2+} load, fractional Ca^{2+} release and diastolic Ca^{2+}.

3.3.2.1 Ca^{2+} kinetics using FLUO-4 dye

$2.5 – 3x10^5$ iPSC-CMs were digested onto Geltrex coated 6 - well plates containing a 25 mm round coverslip. Cardio culture medium was changed every 2 -3 d and after 7 – 10 d cells were used for measurement. CMs were stained with staining solution consisting of 3.8 µl Fluo-4 (1036 µM) mixed vigorously with 0.9 µl Pluronic F-127 and subsequently 1.5 ml 1.8 Tyrode solution was added (2.5 µM Fluo-4 end concentration). CMs were washed once with 1.8 Tyrode solution, then 1.5 ml staining solution was added and CMs were incubated for 30 min at 37 °C in the dark. Staining solution was removed, CMs washed twice with 1.8 Tyrode solution and the coverslip was then removed from the plate and placed into the measuring chamber. 900 µl 1.8 Tyrode solution was added for the measurement. The measurement chamber was positioned on a LSM 720 confocal microscope onto a 63x magnification oil - objective and paced at 0.25 Hz with 18 V/ 3 ms using a Myopacer device. The following settings using the software Zen 2009 were applied: Excitation 488 nm with 1.0 – 2.2 %, a pinhole of 6 AU with 700 gain and offset 0. Measuring lines were drawn in the cytoplasmatic region excluding the nucleus and the following line scan mode was applied: 20.000 cycles, 12 bit unidirectional, zoom factor 3, 512 pixel, no delay and maximum scan speed (945.5 µs per line scan = 18.9 s per total line scan recording). In total 18 – 20 cells/ line scans were recorded for a basal state. Thereafter the 1.8 Tyrode solution was replaced with 1 µM Iso in 1.8 Tyrode solution and cells were allowed to adjust for 10 min before further 18 – 20 line scans were recorded. For measurement with Verapamil, the Verapamil stock solution was pre-diluted with cardio - culture medium from 100 µM to 1 µM to minimize the DMSO content in the measurement. the Tyrode solution for basal measurement was replaced with 1.8 Tyrode solution with 30 nM Verapamil. After the measurement with Verapamil, the solution was replaced again with 1.8 Tyrode containing 30 nM Verapamil and 1 µM Iso.

3.3.2.2 Diastolic Ca^{2+} and Load using FURA –2 dye

3 – 4x10^4 iPSC-CMs were digested onto Geltrex – coated 35 mm round Fluoro dishes. Cardio culture medium was changed every 2 -3 d and after 7 – 10 d, the cells were used for measurement. To load the cells, 5 µl FURA-2 dye was dissolved in 1 ml 1.25 Tyrode solution. The medium in the Fluoro dishes was discarded and the FURA-2 solution added and incubated for 15 min at 37 °C. The FURA-2 solution was discarded and 1.25 Tyrode solution added and incubated for another 15 min at RT. Finally, the 1.25 Tyrode solution was refreshed and the cells measured using a Olympus Motic AE31 microscope with an IonOptix epifluorescence acquisition system. The FURA-2 loaded cells were excited at 340 and 380 nm and the emission was measured at 510 nm. FURA-2 has to property to change its excitation peak from 340 to 380 nm upon Ca^{2+} binding. Therefore, FURA-2 measurements are shown as a 340/380 nm ratio, which gives a direct and comparable Ca^{2+} measurement since it eliminates dye loading differences among the samples. For every Fluoro dish, 3-4 single cells were measured by recording Ca^{2+} transients with increased pacing: 0.25, 0.5, 1.0, 2.0 and 3.0 Hz. The last measured cell was additionally stimulated with caffeine (10 µM) at the diastole, which triggers the complete release of the SR. For this, the 10 mM caffeine stock solution was diluted 1:1000 with 1.25 Tyrode solution. With this, the following parameters were analyzed: diastolic Ca^{2+} content (baseline during diastole), systolic Ca^{2+} content (transient

peak height), SR Ca^{2+}-load (caffeine transient peak height) and fractional Ca^{2+} release (peak height/caffeine peak height).

3.3.3 EdU incorporation assay

EdU is a Thymidine analog carrying an additional ethinylgroup that allows to monitor cells which underwent DNA-synthesis (and therefore mostly mitosis). The incorporated EdU is later visualized by a cycloaddition reaction with a fluorophore. For this, 30 d old 5 – 6x10^4 iPSC-CMs were digested as technical replicates into 3 - 4 wells of a 12 - well plate. 6 to 7 days after digest, the medium was replaced with 1.5 ml cardio culture containing 1 µM EdU solution. After 48 h incubation, the EdU treated cells were fixated. To visualize EdU incorporation, the Click-It EdU Cell Proliferation Kit was used according to the manufacturer's instructions. Additionally, the cells were stained with a cardiac marker titin-Z1/Z2 (1:500) or MLC2v (1:200) to identify CMs. For every differentiation, 3 – 4 coverslips were stained. From every coverslip, 5 – 6 pictures were taken and subsequently the DAPI channel and the EdU channel were counted to calculate a percentage of EdU-positive cells. Cells that were EdU-positive, but negative for the cardiac marker were excluded from analysis.

3.4 STATISTIC and SOFTWARE

3.4.1 Statistics

To test for statistical significances, ANOVA statistics were used when multiple comparisons were required and significant results are marked with * p<0.05, ** p<0.01 and *** p<0.001. In addition, Student's t-test was applied to directly compare two groups (the isogenic lines with the respective patient lines) and significant results are marked with #p<0.05, ## p<0.01 and ### p<0.001.

3.4.2 Software

Different softwares were used for the experiments, which are listed in Tab. 11.

Tab. 11 List of software applications

Experiment	Software
Cell size analysis-2D	ImageJ2/Fiji
EdU analysis	ImageJ2/Fiji
FACS/FLOW	FACS.Diva v7.0
Figures and schematics	MS Power point-2010; Inkscape 1.0
Fluo- Ca^2 sparks analysis	ImageJ2/Fiji (Plugin sparkmaster)
Fluo- Ca^{2+} transients' analysis	ImageJ2/Fiji and Labchart 7.0 (Peak analysis)
FURA- Ca^2 transients' analysis	IonWizard6.4
Gel-imaging for PCR	Image Lab 6.0 (Ethidium bromide setting) and MS Excel-2010
GO term analysis	Cytoscape 3.7 (Plugin ClueGo), WeBgestalt
IF imaging	Axion Biosystems and ImageJ2/Fiji
MEA	MCRack V4.6 (BR, FPD, ISI)
Membrane-imaging (Western blot)	Image Lab 6.0 and MS Excel-2010
Primer design	NCBI primer blast
QPCR	Bio-Rad CFX Maestro 1.1
Microscope imaging	AxioVision Rel4.8
Sarcomere analysis	ImageJ2/Fiji (Plugins: Tubeness, FFT, Radial Profile) and Labchart 7.0 (Peak analysis)
Sequencing alignment	Clustal Omega 3.0
Sequencing FASTA file	Bioedit 6.0
Statistics	GraphPad Prism 8
Ven diagram for DexSeq	Venny2.1

4

RESULTS

4.1 RECRUITMENT OF FAMILIES WITH RBM20-BASED CARDIOMYOPATHIES

For this study of RBM20-based cardiomyopathies, two families with hereditary heart conditions were recruited by the AG Meder at the University hospital Heidelberg. Genomic screening identified different RBM20 mutations as the underlying genetic predisposition. The patients suffer from LVNC or DCM. In both cases of their familial cardiomyopathy, the genetic predisposition was identified in the RBM20 gene affecting the exact same amino acid (aa) position p.R634. This aa position is located in exon 9 and part of a highly conserved RS-domain in RBM20 (Fig 9A), which was identified as a hot-spot for disease-causing point mutations (Li et al., 2010; Brauch et al., 2009). In the LVNC cohort, the RBM20 mutation resembles the heterozygous missense mutation c.G1901T leading to RBM20-p.R634L, whereas in the DCM cohort the heterozygous point mutation c.C1900T translates to RBM20-p.R634W. The family pedigrees and mutations are depicted in Fig. 9B, and the donors of this project are marked (#). The LVNC-causing RBM20 mutation R634L has already been linked to LVNC in a clinical study by the AG Meder (Sedaghat-Hamedani et al.; 2017) and the patients II.3 and III.3 were willing to donate somatic material for this study. As reported, female patient II.3 was diagnosed with LVNC at age 39 and showed with severely reduced ejection fraction of 20 % and non-sustained ventricular tachycardia. Due to these symptoms she received an ICD. Furthermore, an endocardial biopsy revealed fibrosis in her heart tissue (Sedaghat-Hamedani et al., 2017). Similarly, the DCM-linked RBM20 mutation p.R634W has been identified in DCM cohort screenings before, but no further analysis of underlying pathomechanisms was undertaken (Li et al., 2010). The DCM family in this project was not part of a clinal study and has also been identified by the AG Meder. The two patients willing to donate somatic material (III.6 and III.7) are brothers that suffer from severe DCM with reduced ejection fraction of 40 % at 38 years (III.6) and 30 % at 43 years (III.7). In addition, an uncle (II.2) and niece (III.3) were willing to donate somatic material to serve as healthy family controls without mutations in RBM20 and without cardiac symptoms. In conclusion, RBM20 mutations are well known disease variants for LVNC and DCM. The recruitment of two families with different aa substitutions at RBM20 p.634 and different cardiac diseases allows to study the underlying pathomechanisms of RBM20-dependent disease in more detail.

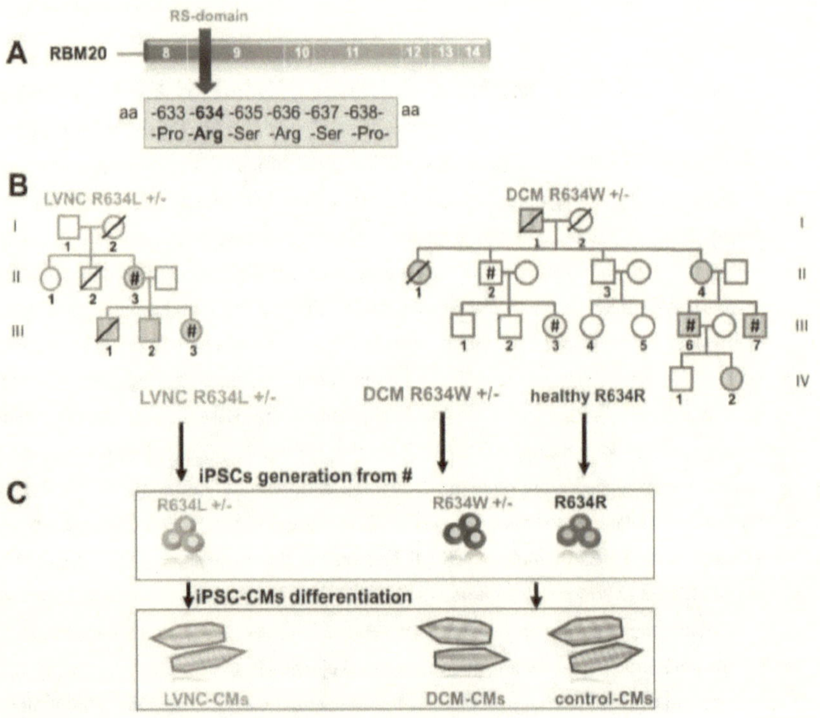

Fig. 9 Families with RBM20-dependent cardiomyopathies
A: Schematic of the RBM20 gene locus with highlighted RS-domain from p.633 till p.638.
B: Families pedigrees of LVNC and DCM. The underlying genetic predisposition in LVNC is RBM20-p.R634L and in DCM RBM20-p.R634W. Symbols colored in grey represent individuals with dysfunctional cardiac conditions. Individuals marked with # donated somatic material for this study.
C: Generation of patient-specific iPSCs and iPSC-CMs. Individuals marked by # in the pedigrees donated somatic material that was reprogrammed into iPSCs and subsequently into iPSC-CMs.

4.1.1 Generation of patient-specific iPSCs

To get further insights into the underlying molecular and cellular mechanisms of RBM20-mutation outcomes, two patients from the LVNC family, two patients from the DCM family and two heathy relatives from the DCM family were willing to donate somatic material (marked with # in Fig. 9B). The donated material consisted either of a skin biopsy (LVNC-II.3, LVNC-III.3, DCM-III.7, healthy-II.2 and III.3) or a blood sample (DCM-III.6), from which fibroblasts or PBMCs were isolated and reprogrammed into iPSCs by the lab technician Johanna Heine, except for DCM-III.6. Reprogramming was carried out by electroporation with OKSM-coding plasmids or Sendai virus (see Tab. 5). This yielded different reprogramming efficiencies of: a) 0.014 % for fibroblasts reprogrammed with plasmids b) 0.01 % for fibroblasts reprogrammed with Sendai virus and c) 0.002 % for PBMCs reprogrammed with Sendai virus. **The generated iPSC – lines are hence forth referred to as LVNC-iPSCs or DCM-iPSCs respectively to make this book more coherent.** The healthy control group consisted of the two relatives from the DCM cohort and five unrelated healthy controls that were already established in the working group (Tab. 5). Notably, the experimental data for the two DCM patients are separated by color in every graph (DCM-III.6 in black; DCM-III.7 in blue). The experimental data for LVNC is derived from the LVNC-II.3 patient since her daughter (LVNC-III.3) joined later during this project and was only included in the NGS transcriptomic analysis.

For every patient, two iPSC-lines were chosen at random for cultivation and characterization for pluripotency. In total, four LVNC iPSC-lines of two patients, four DCM iPSC-lines of two patients, and four control iPSC-lines of two healthy relatives from the DCM family were analyzed from the two cardiomyopathy families. Fig. 10 shows representative pictures of one iPSC-clone for the LVNC (II.3)- and DCM (III.6)-iPSCs. All generated iPSC-lines showed typical stem cell morphology and are positive for alkaline phosphatase activity (Fig 10A). Sanger sequencing of *RBM20*-exon 9 showed the expected missense mutation for LVNC and DCM (Fig 10B). Semi-quantitative PCR demonstrated upregulation of pluripotency markers *OCT4, SOX2, NANOG, LIN28* and *GDF3* in all iPSC-lines. Somatic material like fibroblasts or PMBCs, which were used for reprogramming are negative for these markers (Fig 10C). Furthermore, expression of pluripotency-linked proteins can be verified for all iPSC-lines demonstrated by immunofluorescence stainings against OCT4, SOX2, NANOG, LIN28, TRA1-60 and SSEA4 (Fig 10D). To test their differentiation capacity, the iPSCs were cultured as embryoid bodies (EBs) to induce spontaneous differentiation. Subsequently, the EBs were analyzed for the three germlayers demonstrated by mRNA expression for albumin (*ALB*) and α-fetoprotein (*AFP*), (endodermal markers), α-*MHC* (mesodermal marker), and microtubule-associated protein 2

(*MAP2*) (ectodermal marker) (Fig 10E). In contrast, pluripotency markers like *OCT4* and *NANOG* became downregulated during EB formation (Fig 10E). Protein expression of the three germlayers was verified by IF for AFP, mesodermal α-smooth muscle actin (α-SMA), and ectodermal βIII-tubulin (Fig 10F).

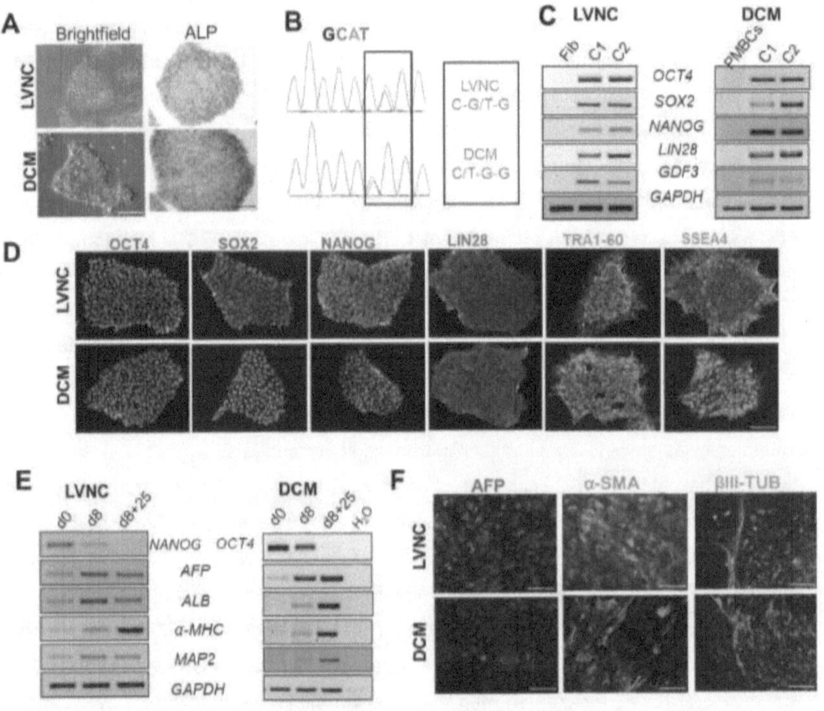

Fig. 10 Characterization of patient-specific iPSCs and iPSC-CMs from LVNC and DCM
A: Brightfield and ALP images of LVNC and DCM-iPSCs (scale bar = 200 µm), ALP= alkaline phosphatase
B: Sanger sequencing of exon 9 from RBM20 locus.
C: Semi-quantitative PCR for pluripotency markers *OCT4, SOX2, NANOG, LIN28* and *GDF3* in LVNC- and DCM-iPSCs.
D: Immunofluorescence stainings for pluripotency markers OCT4, SOX2, NANOG, LIN28, TRA1-60 and SSEA4 (scale bar = 100 µm) in LVNC- and DCM-iPSCs.
E: Semi-quantitative PCR for germlayer markers in EBs: *AFP, ALB, α-MYH* and *MAP2*. Pluripotency markers *OCT4* and *NANOG* become downregulated in LVNC- and DCM-EBs.
F: Immunofluorescence stainings for germlayer markers: AFP, α-SMA, βIII-TUB (scale bars = 100 µm) in LVNC- and DCM-EBs.

4.1.2 Directed differentiation into iPSC-CMs

To study RBM20-dependent cardiac diseases the iPSCs were differentiated into functional beating CMs. The differentiation protocol is adapted from Lian et al. and utilizes small molecules to activate and subsequently inhibit Wnt signalling (Lian et al., 2013) (Fig. 11A). A metabolic selection step was included during which glucose is replaced by lactate as a carbon source. This is feasible to dispose of glucose-dependent non-cardiomyocytes which may have formed during the differentiation (Tohyama et al., 2013). Subsequently, the iPSC-CMs were cultured for >60d since prolonged culture duration enhances the cardiac maturation. This protocol is feasible to differentiate beating ventricular CMs, which expressed cTNT and showed sarcomeric structure (Fig. 11B). To monitor the differentiation efficiency and purity, the cells were quantified in FLOW analysis after staining with the cardiac marker cTNT. Cardiac differentiation purity was comparable among all iPSC-lines with 91 % ± 6 cTNT positive cells (Fig. 11C). In summary, the cardiac differentiation of iPSCs into iPSC-CMs was successful. As described in the preceding section, two iPSC-lines were derived for every patient of which multiple cardiac differentiations were used for the following experiments. The number of differentiations is indicated with n for every graph. The differentiation into iPSC-CMs serves as a human *in vitro* model to study RBM20-dependent LVNC and DCM.

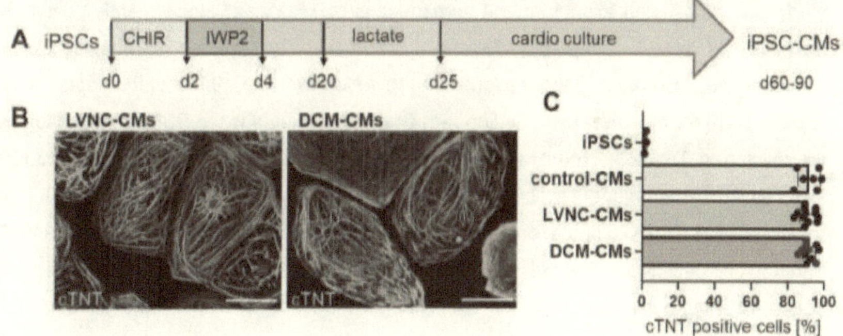

Fig. 11 Cardiac differentiation of iPSCs
A: Protocol for iPSC-CMs differentiation. Differentiation start is marked as d0 with addition of CHIR (CHIR99021), followed by addition of IWP2. Protocol adapted from (Lian et al., 2013; Tohyama et al., 2013).
B: Immunofluorescence stainings for the cardiac marker cTNT in LVNC- and DCM-CMs (scale bars = 50 µm).
C: Analysis of differentiation efficiency by cTNT-FLOW for control-, LVNC- and DCM-CMs. Every dot represents one cardiac differentiation. Undifferentiated iPSCs serve as negative control.

4.2 ANALYSIS of RBM20-DEPENDENT SPLICING

4.2.1 Establishment of RBM20 splice targets at the human locus

In 2012, Guo and colleagues published RBM20 splice targets based on a knockout rat model. They pointed out that splicing of these genes is conserved among species and can be transferred as RBM20-dependent splice targets in the human system. These splice targets included the essential cardiac genes *TTN*, *RYR2*, *CAMK2γ*, *CAMK2δ*, triadin (*TRDN*), lim-domain binding protein 3 (*LDB3*), sorbin and SH3 domain containing 1 (*SORBS1*) and the Ca^{2+} voltage-gated channel subunit alpha 1 (*CACNA1c*) (Guo et al., 2012). To study if splicing in these genes is also affected in the human LVNC- and DCM-CMs harboring the described RBM20 mutations, these splicing targets were translated to the human loci. Therefore, all rat-based primers were blasted and adjusted against the human gene locus of the respective gene and semi-quantitative PCR was performed to amplify the exons of interest. When the PCR products were visualized on an agarose-gel, multiple bands could be detected corresponding to different splice products of the gene. These bands were cut from the gel and subjected to Sanger sequencing to identify, which exons are included in the PCR product of a gene. *TRDN* is depicted as an example for this approach in Fig. 12A. For *TRDN*, the primer blast revealed binding in exon 7 and exon 10 in the human locus of *TRDN* and subsequently primers were designed for exon 7 (forward) and exon 10 (reverse). The agarose gel showed two PCR products and subsequent Sanger sequencing revealed exon 7, 8, 9, 10 for the upper band and exon 7, 8, 10 for the lower band. In line with Guo et al., exon 9 is the RBM20-dependent exon for splicing in *TRDN*. With this approach the following exons were identified for the human locus as RBM20-dependent splice targets: exon 5 in *LDB3*, exon 5 in *SORBS1*, exon14 (coding for an NLS) in *CAMK2δ* and exon 14 (coding for an NLS) in *CAMK2γ*. Notably for *CAMK2δ*, 3 bands were detected in the gel, whereas only the middle and lower band was successfully sequenced and corresponded to the *CAMK2δ* gene (Fig. 12B). Screening of *CACNA1C* only showed one PCR product (data not shown), but primers were designed for exon 9 as this is the reported RBM20-dependent exon for splicing (Guo et al., 2012). *TTN* and *RYR2* splicing were not verified by this approach because they were already published for the human locus: N2A domain in *TTN* (Streckfuss-Bömeke et al., 2017) and a cryptic 24-bp insertion from Intron 80-81 in *RYR2* (George et al., 2007).

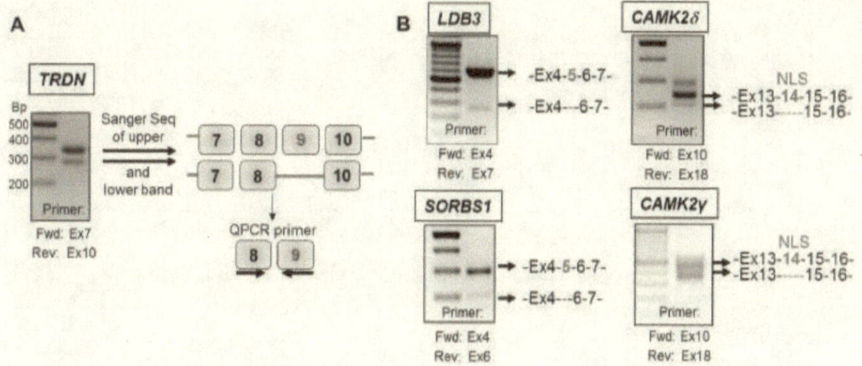

Fig. 12 Adaption of RBM20 splice-targets to the human locus
A: Example for establishment of RBM20-exon specific QPCR-primer of the splice target *TRDN*.
B: Sanger Sequencing of different PCR products revealed the RBM20-specific exon for splicing in target genes (marked in red).

4.2.2 RBM20 splice-targets are affected differentially in LVNC-CMs and DCM-CMs

The RBM20-dependent splice targets that were adjusted for the human loci, were subjected in quantitative PCR to assess how splicing is affected in LVNC- and DCM-CMs. Splicing of *TTN* and *RYR2* revealed RBM20-dependent missplicing in LVNC- and DCM-CMs with *TTN*-mRNA that includes the N2BA and *RYR2*-mRNA that includes a cryptic 24bp insertion from intron 80-81 (Fig. 13A). Intriguingly, inclusion of exon 9 in *TRDN* and an NLS (exon 14) in *CAMK2δ* is only affected in LVNC-CMs, but not in DCM-CMs compared to control-CMs. In contrast, splicing of exon 5 in *LDB3* is affected in DCM-CMs exclusively (Fig. 13B). Notably, the overall gene expression of these splice targets was not affected underscoring the influence of RBM20 in splicing and not gene expression (Fig. 13C). For *TTN*, the altered N2BA/N2B ratios observed in LVNC- and DCM-CMs were due to significantly decreased levels of N2B (Fig. 13C). Furthermore, other tested RBM20 splice targets were not found to be affected by the RBM20 mutation in LVNC- and DCM-CMs such as an NLS in *CAMK2γ*, exon 5 in *SORBS1*, exon 9 in *CACNA1c and exon 15/16 in CAMK2δ* (Fig. 13D). In order to analyze whether the amount of RBM20 itself is regulated, the mRNA levels of *RBM20* were analyzed. QPCR-results showed that *RBM20* levels in LVNC-CMs are significantly decreased compared to control- and DCM-CMs (Fig. 13E).

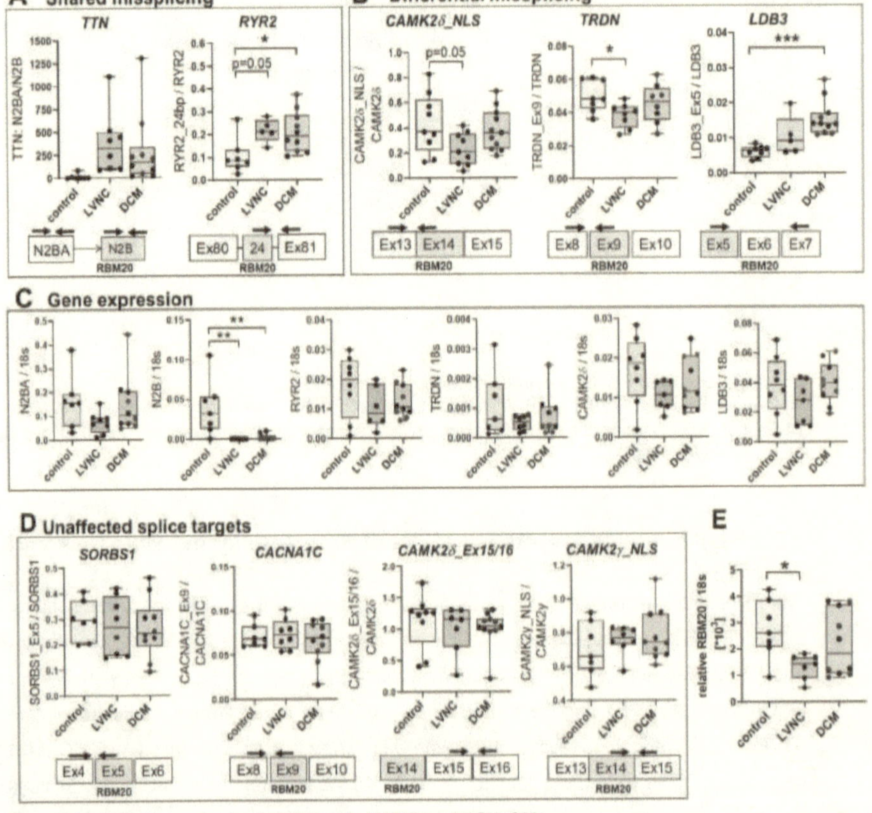

Fig. 13 RBM20-dependent missplicing in LVNC and DCM-CMs

A-E: Every dot represents 1 cardiac differentiation. Data is presented as boxplots. P-values are determined by one-way ANOVA against control with Dunnett correction and results are marked with * p<0.05, ** p<0.01 and *** p<0.001.

A,B,D: Schematics under each graph depict the RBM20-dependent exons and introns (in yellow). Arrows indicate the positions of the primers used for every gene.

A: RBM20 splice-targets *TTN* and *RYR2* are misspliced in LVNC- and DCM-CMs.

B: RBM20 splice-targets *TRDN* and *CAMK2δ-exon14* are affected in LVNC-CMs exclusively. *LDB3* is affected in DCM-CMs.

C: The overall expression of RBM20 splice targets is not affected.

D: RBM20 splice-targets, *SORBS1, CACNA1C, CAMK2γ* and CAMK2δ-exon 15/16 are not affected in LVNC- and DCM-CMs.

E: *RBM20* mRNA is downregulated in LVNC-CMs.

In conclusion, this assay revealed that RBM20-dependent splice targets identified by knockout rats can be transferred to the human locus and furthermore identified shared and differential missplicing events in LVNC- and DCM-CMs. Notably, the splicing of *TRDN* and *LDB3* displayed converse inclusion/exclusion results. In RBM20 knockout rat, *TRDN*-exon 9 levels were increased and *LDB3*-exon 5 levels were decreased (Guo et al., 2012). In this study of human iPSC-CMs, the *TRDN*-exon 9 levels were decreased, and *LDB3*-exon 5 levels increased. Whether this deviation is due to inter-species differences or an effect of overactive RBM20-dependent splicing remains to be determined. The differential missplicing could be one molecular basis for the development of different cardiac diseases that are both based on an RBM20 mutation. Decreased *RBM20* mRNA levels in LVNC-CMs suggest a haploid insufficiency mechanism to contribute to the disease.

4.3 CELLULAR and PHYSIOLOGICAL INVESTIGATION OF LVNC- and DCM-CMs

RBM20 splice targets comprise cardiac genes that contribute to the sarcomere structure (*TTN, LDB3*), Ca^{2+} homeostasis (*RYR2, CAMK2δ, TRDN*) and electrical activity (*CACNA1C*). Therefore, LVNC- and DCM-CMs were tested on these physiological parameters. In the next set of experiments, multiple CMs were measured derived from multiple cardiac differentiations. Therefore, the analysis was calculated by nested-statistic to take the cardiac differentiations into account. In addition, statistical analysis was also performed with Student´s t-Test and ANOVA calculations without taking the differentiations into account. The latter is provided in the supplements where the graphs depict every data point for the respective experiment.

4.3.1 Sarcomeric integrity is decreased in LVNC- and DCM-CMs

A major function of cardiomyocytes is the generation of force by contraction of the cells. This is possible because of a highly regular, structured and contractile cytoskeleton that is composed of sarcomeres. The succession of many sarcomeres within a cell arrange into a myofibril, which arrange into a muscle fiber. The sarcomeres represent the smallest contractile unit in a muscle cell and require the correct assembly of specialized proteins such as actin, myosin, titin, troponin and many other structural and regulatory proteins (Gautel & Djinović-Carugo, 2016). Since RBM20 is involved in splicing of sarcomeric proteins such as titin, the sarcomeric integrity was analyzed by immunofluorescence. A double staining was used to visualize the sarcomeric Z-disc with α-actinin and the M-line with titin (M8/M9) antibody followed by quantification of sarcomere regularity using fast Fourier Transformation (FFT) (Suppl. Fig. 1A). The double staining unveiled a disturbed sarcomeric structure for LVNC- and DCM-CMs with areas that lack any repeated sarcomere organization (Fig. 14A). The quantification coincides with this observation and showed that the sarcomeric regularity is significantly decreased 1.6-fold in LVNC and DCM-CMs for the Z-disc (Fig. 14B) and 1.3-fold for the M-line (Fig. 14C) compared to control-CMs. This is in line with the observation of *TTN* missplicing in LVNC- and DCM-CMs and underscores the importance of RBM20 for sarcomeric regularity.

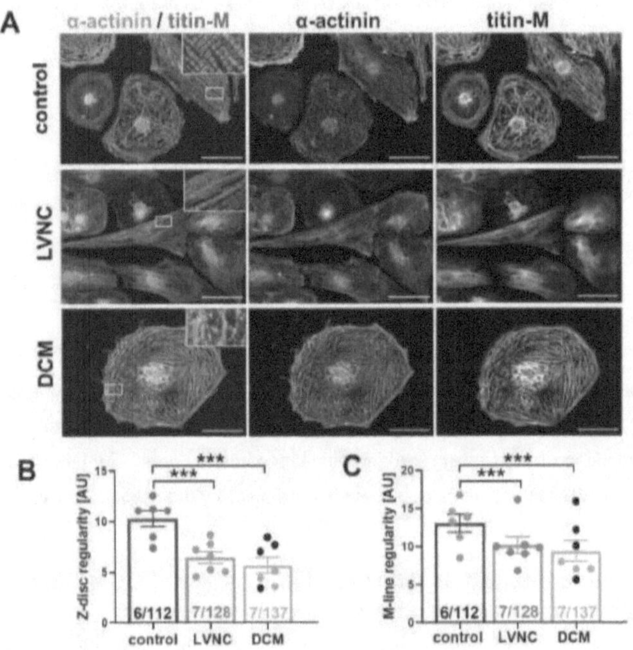

Fig. 14 Sarcomeric regularity is decreased in LVNC- and DCM-CMs
A: Visualization of the sarcomere by immunofluorescence stainings against α-actinin (green) and titin M8/M9 (red) for control-, LVNC- and DCM-CMs (scale bars = 50 µm).
B + C: Quantification of Z-disc (**B**) and M-line (**C**) sarcomeric regularity using fast Fourier transformation. Bar graphs depict the peak amplitude of the first-order peak. Multiple pictures were analyzed [No. of differentiations/analyzed pictures] are indicated in the graph. Data represented as nested mean +/-SEM. P-values were determined by nested one-way ANOVA multiple comparisons with Dunnett correction and results are marked with * p<0.05, ** p<0.01 and *** p<0.001.

4.3.2 Sarcomeric dysregulation is a common phenotype in splice-defect associated DCM

An irregular and disorganized sarcomere has been shown before in other cardiomyopathies including another RBM20 mutation-based DCM, in which affected patients carry the heterozygous missense mutation p.S635A (Streckfuss-Bömeke et al., 2017). Therefore, the question was addressed if sarcomeric irregularity is a common characteristic of splice-defect associated cardiomyopathies. For this, the following iPSC-CMs were analyzed: 1) A family with hereditary DCM carrying heterozygous double mutations in SCN5A (p.C335R) and RBFOX1 (p.S107F) or in the SCN5A (p.C335R) only. 2) A family with hereditary DCM harboring a genetic predisposition in the splice-substrate LMNA with two intronic variants (rs199686967 and rs201379016) and a synonymous mutation (rs149339264). Both families were identified by the AG Meder in Heidelberg. 3) A newly identified splice factor of cardiac structural proteins, SLM2, was knocked down in control-CMs. As seen in Fig. 15A and B, sarcomeric regularity was decreased for all iPSC-CMs if a splice factor or substrate is affected. In the RBFOX1/SCN5A cohort, sarcomeric integrity was only impaired if the splice factor RBFOX1 (Frese et al., 2015) harbors the variant p.S107F. IPSC-CMs from patients that solely carry the mutation in the ion channel SCN5A (p.C335R) did not show a decrease in sarcomere regularity (Fig. 15A). The *LMNA* gene gives rise to two isoforms Lamin A and Lamin C and is a structural protein that forms the nuclear envelope. Patients with mutations in *LMNA*, also collectively termed laminopathies, often present with symptoms including cardiac diseases (Boriani et al., 2015). SLM2 is a ubiquitous expressed splice factor that has splice targets in the heart including *TTN*. To understand the contribution of SLM2 in sarcomeric organization, AAV6 mediated knockdown was accomplished in iPSC-CMs. This was part of a SLM2 characterization project in corporation with the AG Boeckel in Leipzig. Because the AAV6-shRNA-SLM2 vector was tagged with GFP, the Z-disc was visualized with titin Z1/Z2 antibody instead of α-actinin. In both entities, *LMNA* mutation and *SLM2* knockdown, the sarcomere regularity is significantly decreased (Fig. 15A and B). In summary, this adds to the evidence that disturbed sarcomeric integrity is a major factor and common trait in splice-defect associated cardiomyopathies that contributes to the disease phenotype.

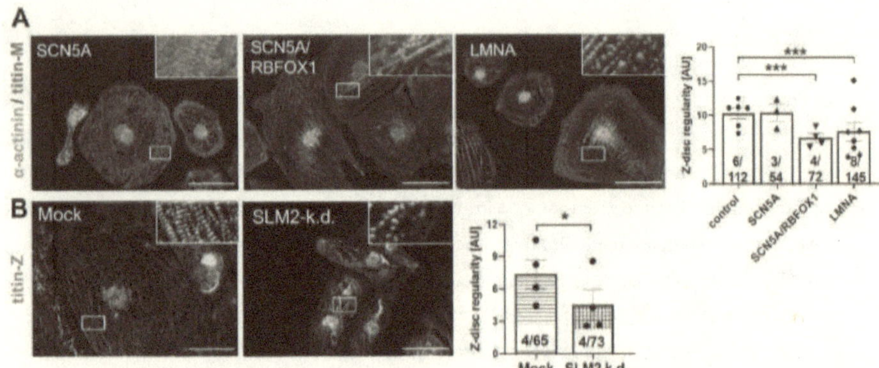

Fig. 15 Sarcomeric regularity is decreased in other mutation-based DCM-CMs
A+B Multiple pictures were analyzed [n = number of differentiations/analyzed pictures] and are indicated in the graph. Data represented as nested mean +/-SEM. P-values were determined by nested one-way ANOVA multiple comparisons with Dunnett correction and results are marked with * $p<0.05$, ** $p<0.01$ and *** $p<0.001$. Scale bars = 50 µm

4.3.3 Ca²⁺ handling impairments in LVNC- and DCM-CMs

Correct Ca^{2+} cycling is a crucial event in muscle cells as Ca^{2+} translates the electrical stimulus into the mechanical contraction of the cell. This process requires a delicate interplay of Ca^{2+} handling proteins whereas impairments in Ca^{2+} homeostasis have been shown numerous times as a major pathology in cardiac diseases (Marks, 2013). Since RBM20 splice targets comprise a set of important Ca^{2+} handling players including *RYR2, TRDN* and *CAMK2δ*, Ca^{2+} homeostasis was investigated in in LVNC- and DCM-CMs. *TRDN* and *CAMK2δ* missplicing was only observed in LVNC-CMs suggesting an altered Ca^{2+} phenotype in LVNC-CMs. To test this hypothesis, Ca^{2+} homeostasis was analyzed in the iPSC-CMs.

4.3.3.1 Ca²⁺ cycling is accelerated in LVNC-CMs

The iPSC-CMs were loaded with the cytoplasmatic Ca^{2+}-probe Fluo-4, which makes it possible to visualize Ca^{2+} transients and determine rise and decay time as well as the Ca^{2+} leakage. Fig. 16A depicts examples of Ca^{2+} transients at basal and after Iso stimulation. First, Ca^{2+} cycling rise and decay time were quantified. The recorded transients displayed an average rise time of 330 ms (SD=156.6) and decay time of 867 ms (SD=245.2) for control-CMs. In comparison, LVNC- and DCM-CMs show significantly faster cycling rates with reduced Ca^{2+} transient rise times of 212 ms (SD=100) for LVNC-CMs and 257 ms (SD=130) for DCM-CMs and Ca^{2+} transient decay times of 626 ms (SD=220) for LVNC-CMs and 739 ms (SD=307) for DCM-CMs. Notably, the LVNC-CMs demonstrated the most prominent decrease in Ca^{2+} cycling times that are also significantly reduced compared to DCM-CMs. To further investigate the Ca^{2+} cycling rates, the cells were incubated with Iso. Iso stimulates the ß-adrenergic pathway and facilitates Ca^{2+} release and reuptake, thereby leading to faster cycling dynamics. Control-CMs showed a physiological reaction to Iso stimulation by a significant decrease of Ca^{2+} transient rise time to 242 ms (SD=138) and decay time to 687 ms (SD=235). DCM-CMs also exhibited this Iso-dependent effect with a significant decrease in rise time of 202 ms (117=SD) and decay time of 522 ms (SD=252). Intriguingly, LVNC-CMs did not show a prominent response to the incubation with Iso and demonstrated similar transient rise times (Basal:Iso 212:206 ms) and only a weak reaction in the decrease of decay time (Basal:Iso 626:540 ms) possibly due to their already fast Ca^{2+} cycling at basal level (Fig. 16B). Furthermore, Ca^{2+} sparks were assessed as a parameter for pro-arrhythmic tendencies. Ca^{2+} sparks are spontaneous openings of RYR2 during the diastole and are also visualized by the Fluo-4 probe (Fig. 16C). The spark frequency is comparable among all iPSC-CMs (Fig. 16D), whereas the Ca^{2+} leakage is significantly increased for DCM-CMs (Fig. 16E). The Ca^{2+} leakage is defined by the sum and size of all sparks, therefore DCM-CMs did not exhibit more, but bigger Ca^{2+} sparks. In conclusion, this is the next evidence that LVNC-CMs and DCM-CMs with their underlying RBM20 mutation showed different disease phenotypes. LVNC-CMs exhibited accelerated Ca^{2+} kinetics and an impaired reaction to Iso. In contrast, DCM-CMs demonstrated faster Ca^{2+} kinetics as well, but a physiological reaction to Iso and revealed a significant higher Ca^{2+} leakage compared to control- and LVNC-CMs.

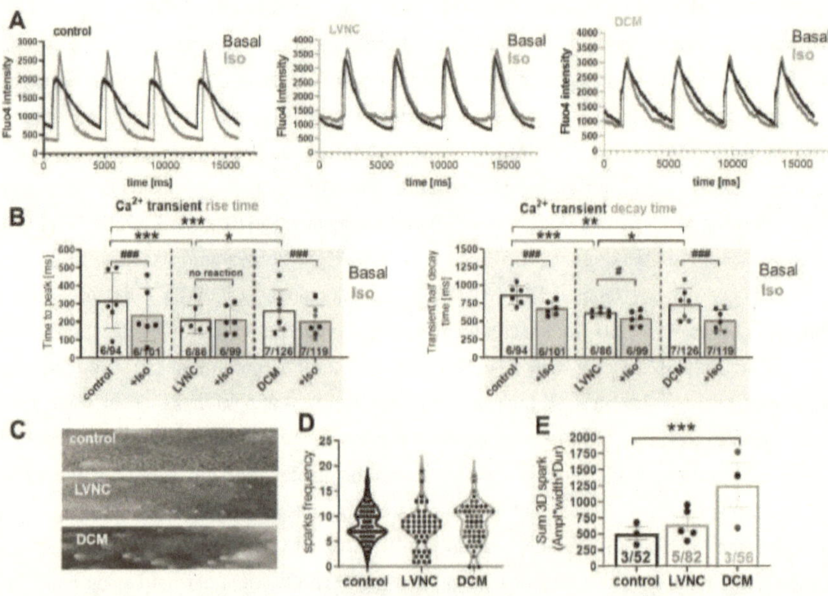

Fig. 16 LVNC- and DCM-CMs show pathologies in Ca^{2+} handling.
A: Exemplary single-cell Ca^{2+} traces from iPSC-CMs of control, LVNC and DCM at basal (black) and Isoprenaline (Iso) (green) stimulation (1 µM) measured using Fluo-4. The cells were paced at 0.25 Hz.
B: Ca^{2+} transient rise (left) and decay time to 50 % (right) for basal and Iso-stimulated (1 µM) condition. LVNC- and DCM-CMs show decreased basal rise and decay time levels, whereas LVNC-CMs exhibit the most decrease. Additionally, LVNC-CMs show a weak reaction to Iso stimulation. Data is presented as nested mean+/- SEM; p-values were calculated by nested one-way ANOVA against control of the basal condition with Dunnett correction and results are marked with * p<0.05, ** p<0.01 and *** p<0.001. Additionally, #p-values were calculated for basal und Iso-treated (1 µM) condition for each group using nested Student's t-Test and results are marked with # p<0.05, ## p<0.01 and *### p<0.001. Multiple cells were analyzed [No. of differentiations/analyzed pictures] and are indicated in the graphs.
C: Representative images of Ca^{2+} spark detection during diastole using Fluo-4.
D: Analysis of Ca^{2+} spark frequency. Multiple cells were analyzed [No. of differentiations/analyzed pictures] and are indicated in the graph. Data is presented as Violin plot for every measurement.
E: Analysis of Ca^{2+} sparks showed increased Ca^{2+} leakage for DCM-CMs using the ImageJ Sparkmaster plugin. Quantitative calculation of 3D Ca^{2+} leakage at for: [No. of differentiations/ analyzed cells] are indicated in the graph. Data is presented as nested mean+- SEM; p-values were calculated by nested one-way ANOVA multiple comparisons of the basal condition with Dunnett correction and results are marked with * p<0.05, ** p<0.01 and *** p<0.001.

4.3.3.2 Systolic Ca²⁺ is increased in LVNC-CMs and decreased in DCM-CMs

To complement this data, single iPSC-CMs were stained with a FURA-2 probe to investigate diastolic Ca^{2+}, Ca^{2+}transient amplitude at the systole, SR-load and fractional Ca^{2+} release. This is possible because FURA-2 is a ratiometric dye that shifts its emission spectrum from 380 nm (FURA-2) to 340 nm (FURA-2 bound to Ca^{2+}) and therefore eliminates the problem of probe-loading because the value is presented as a 340/380 nm ratio. Ca^{2+} transients were recorded continuously with 0.25 Hz pacing and lastly caffeine was added, which triggers a fast release of the stored Ca^{2+} from the SR. Diastolic Ca^{2+} levels were decreased in both, LVNC- and DCM-CMs, compared to control-CMs (Fig. 17A). Intriguingly, LVNC-CMs exhibit increased and DCM-CMs decreased systolic Ca^{2+} amplitudes (Fig. 17B). This parameter is an indicator for force generation, as more Ca^{2+} correlates with stronger contraction thereby indicating that DCM-CMs showed weaker contraction strength. The SR load measures the Ca^{2+} content in the SR, whereas the SR fractional release is a parameter that shows how much Ca^{2+} gets extruded from the SR during a contraction. In this study, the SR load and SR fractional release was comparable between control-, LVNC- and DCM-CMs (Fig. 17C and D). Taken together this data demonstrates that the two RBM20-mutation based cardiomyopathies exhibited different Ca^{2+} handling impairments that possibly contribute to the different disease phenotypes. LVNC-CMs showed faster Ca^{2+} cycling and cannot react upon β-adrenergic stimulation with Iso. In contrast, DCM-CMs showed a physiological reaction to Iso but exhibited an increased pro-arrhythmic Ca^{2+} leak and decreased systolic Ca^{2+}.

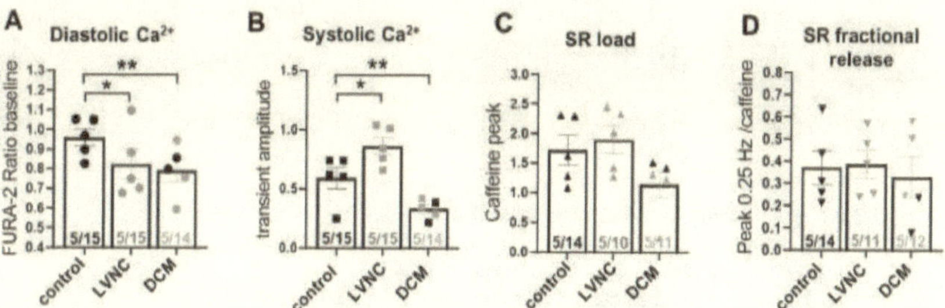

Fig. 17 Systolic and diastolic Ca²⁺ is impaired in LVNC- and DCM-CMs
A-D: Ca^{2+} parameters measured by the ratiometric FURA-2-am probe in single cell iPSC-CMs. Data is presented as nested mean+/- SEM and p-values were calculated by nested one-way ANOVA against control of the basal condition with Dunnett correction and results are marked with * p<0.05, ** p<0.01 and *** p<0.001. Number in the graph represent [No. of differentiations/measured cells].
A: Diastolic Ca^{2+} is reduced in LVNC- and DCM-CMs compared to control-CMs.
B: Systolic Ca^{2+} increased in LVNC-CMs and decreased in DCM-CMs compared to control-CMs.
C: SR- Ca^{2+} load is unaffected in LVNC- and DCM-CMs compared to control-CMs.
D: SR fractional Ca^{2+} release is unaffected in LVNC- and DCM-CMs compared to control-CMs.

4.3.4 Electrophysiological properties in cardiomyocytes

To complement the set of physiological assessments, LVNC- and DCM-CMs were analyzed electrophysiologically by multi-electrode array (MEA). In the MEA experiments, the iPSC-CMs were digested onto small electrodes recording extracellular field potentials that show the Na^+ spike and the t-wave (Fig. 18A). With this, the beating rate (BR), field potential duration (FPD) and interspike interval (ISI) can be analyzed (Fig. 18A). Upon addition of Iso (100 nM), a reaction in form of an increased beating rate was observed (Fig. 18B). It has been reported that RBM20 splice targets comprise important voltage-sensitive channels like *CACNA1C* and that patients with RBM20 mutation-based DCM often suffer from arrhythmic events (van den Hoogenhof et al., 2018). LVNC- and DCM-CMs did not differ significantly from control-CMs with an average basal BR of 0.7 Hz and cFPD of 250 ms (Fig. 18C and D). Upon stimulation with Iso, all iPSC-CMs showed a comparable 2-fold increase in BR and a subsequent recovery when Iso was removed (Fig. 18E). Analysis of the ISI revealed no irregular beating for LVNC- and DCM-CMs, comparable to the stable beating of control-CMs (Fig. 18F).

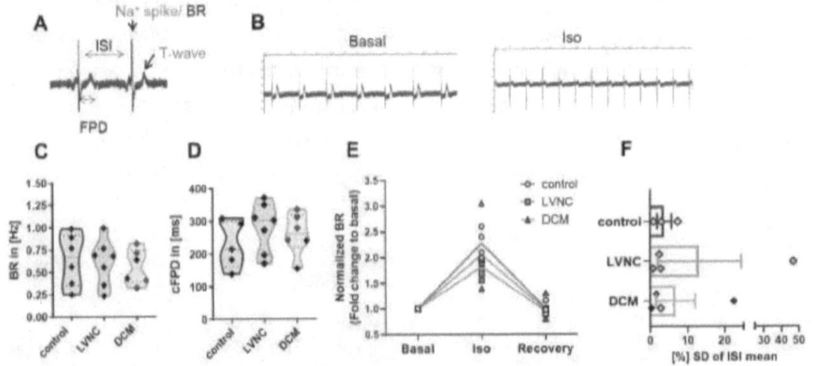

Fig. 18 Electrical activity in LVNC- and DCM-CMs is comparable to control-CMs.
A: MEA recordings show a Na^+ spike and the t-wave, which is used for calculation of the beating rate (BR), field potential duration (FPD) and interspike interval (ISI).
B: Example of accelerated electrical activity after Iso (100 nM) stimulation.
C-F: Every dot represents one cardiac differentiation. Data was tested with one-way ANOVA against control with Dunnett correction, but no significances were detected.
C: Beating rate is comparable among control-, LVNC-, and DCM-CMs.
D: Increase of ca. 2-fold in BR after Iso (100 nM) is added. Subsequently Iso is removed and CMs were re-measured after 24 h. Recovery rate is comparable to basal rates.
E: Frequency-corrected FPD (cFDP) shows comparable values among control-, LVNC-, and DCM-CMs.
F: For quantification of beating regularity, the percentage of StDev to the mean of ISI was calculated. Except for two cardiac differentiations in LVNC and DCM, all CMs show regular beating.

Taken together, BR and cFDP were comparable among control-, LVNC- and DCM-CMs with a prominent reaction to Iso. In the Ca^{2+} analysis DCM-CMs showed increased Ca^{2+} leakage that can contribute to arrhythmic events. However, based on the MEA analysis, no significant arrhythmic events were detected.

4.4 RBM20 GENE EDITING

4.4.1 Generation of RBM20-rescue lines using CRISPR/Cas9 technology

The data demonstrated that distinct RBM20 mutations at the same aa position 634 can segregate into different disease phenotypes, especially in Ca^{2+} handling (Fig.17 and 18). To verify that solely the distinct RBM20 variant is the causative mutation, isogenic rescue-RBM20 lines were generated by using the CRISPR/Cas9 technology. Therefore, one LVNC (from LVNC-II.3) and one DCM-iPSC-line (from DCM-III.6) were chosen and the heterozygous missense mutation was converted back into wt RBM20 expressing arginine at aa position 634 homozygously (Fig. 19A). For this purpose, crRNAs were designed to cut close to the SNP-mutation and the HDR template was designed to code for wt RBM20 (see Fig. 7). 72 clones for each, rescue-LVNC and rescue-DCM, were picked and subsequently screened for the desired gene editing. Rescue-DCM yielded 2.8 % successful wt RBM20 editing (2/72 edited clones) and rescue-LVNC 45 % successful editing (14/31 edited clones). Sequencing of the edited clones for the RBM20 locus furthermore revealed that not every edited clone contained the artificial silent mutations, which were coded on the HDR template (Fig. 19B). For example, resDCM1 showed only edited wt RBM20 without any further nucleotide substitutions, indicating that the other allele was used for HDR instead of the exogenously provided HDR template. Notably, all generated rescue-lines were used, regardless whether a silent mutation was included. Two rescue-DCM-iPSC lines and four randomly chosen rescue-LVNC-iPSC lines were cultivated and characterized for expression of pluripotency markers and CM-differentiation efficiency. Furthermore, RBM20 was sequenced (Fig. 19B) every 5 passages to assure RBM20 correction.

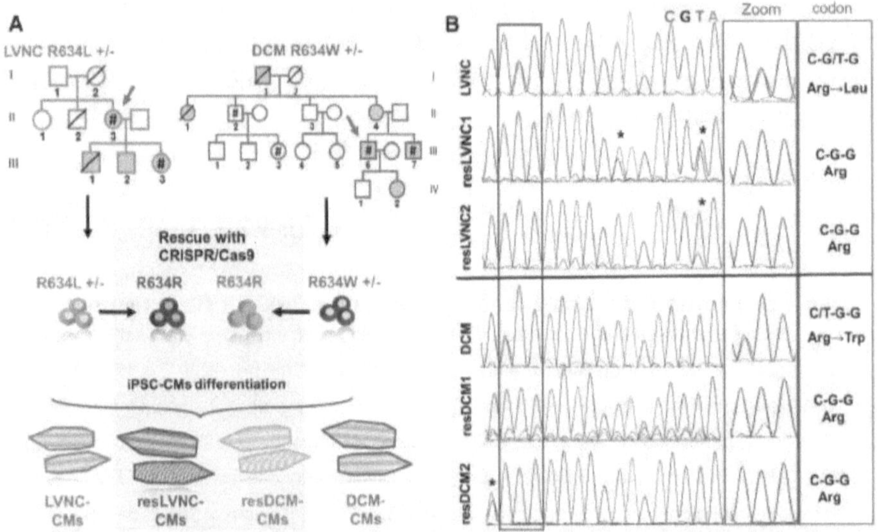

Fig. 19 Generation isogenic resLVNC- and resDCM-iPSCs
A: Schematic of generation of isogenic RBM20-rescue lines from LVNC- and DCM-iPSCs. The red arrows indicate the patient of which gene editing was performed.
B: Sanger sequencing results of RBM20 exon 9. The zoom visualizes position c.1900-1902 which encodes aa 634 within the RBM20 gene. Wt-RBM20, which encodes C-G-G for p.R634, is restored in resLVNC- and resDCM-iPSCs. * marks non-synonymous mutations included by CRISPR/Cas9 editing.

Similar to the patient iPSC-lines, the rescue-RBM20 lines expressed pluripotency-related proteins OCT4, SOX2, NANOG, LIN28, GDF3, FOXD3, SSEA4 and TRA1-60 (Fig. 20A) and showed spontaneous differentiation into cells of the endo- meso- and ectoderm (Fig. 20B). Analysis of cTNT expression by FLOW showed a high CM differentiation purity of 89 to 96 % (Fig. 20C). It can be concluded that the generation of rescue-RBM20-iPSC and further rescue-RBM20-CMs was successful for LVNC and DCM and these edited iPSCs showed no difference in their pluripotency characteristics or their iPSC-CMs differentiation potential. **Therefore, these edited iPSC-lines can serve as isogenic controls to LVNC and DCM and are referred to as resLVNC (rescue of RBM20 mutation in LVNC) and resDCM (rescue of RBM20 mutation in DCM).** Notably, the DCM- and resDCM-iPSCs were published by our group in the course of this project (Rebs et al., 2020). As mentioned earlier, the two DCM patients are displayed in different colors in every graph. Since the DCM-III.6 patient was represented with black symbols, the data of resDCM (gene editing of DCM-III.6) is also represented as black symbols.

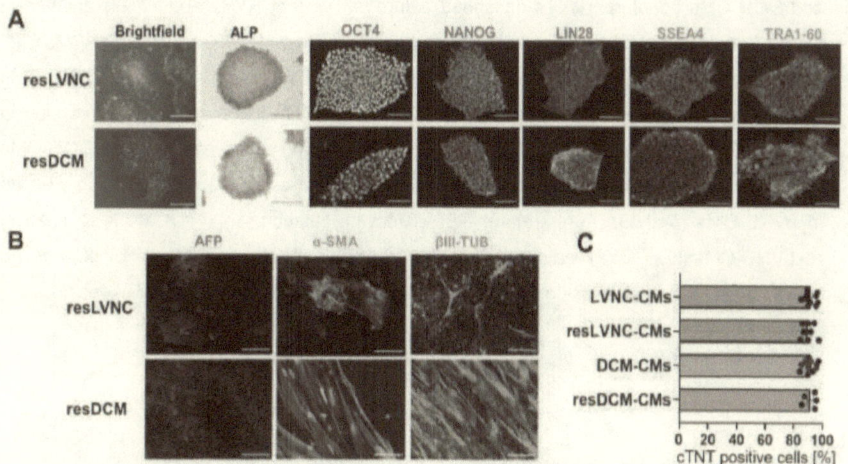

Fig. 20 Characterization of isogenic resLVNC- and resDCM-iPSCs
A: Microscope images for resLVNC- and resDCM-iPSCs. Both iPSC-lines show typical stem cell morphology and activity of ALP (scale bars = 200 µm). Immunofluorescence staining are positive for pluripotency markers OCT4, NANOG, LIN28, SSEA4 and TRA1-60 (scale bars = 100 µm). ALP: Alkaline phosphatase
B: Immunofluorescence stainings for germlayer markers: AFP, α-SMA, βIII-TUB (scale bars = 100 µm) in resLVNC- and resDCM-EBs.
C: Comparison of differentiation efficiency by cTNT-FLOW in LVNC-, resLVNC-, DCM- and resDCM-CMs. Every dot represents one cardiac differentiation. Undifferentiated iPSCs serve as negative control.

4.4.2 RBM20-rescue lines confirmed the shared and differential disease properties in LVNC- and DCM-CMs

The generated rescue RBM20-lines were differentiated into CMs and splicing, sarcomeric integrity, Ca^{2+} kinetics and electrical activity were reassessed to compare to their patient LVNC- and DCM-CMs counterparts. As described before, splicing of *RYR2* and *TTN* is affected in LVNC- and DCM-CMs and congruently splicing is restored in resLVNC- and resDCM-CMs (Fig. 21A). The *TRDN*-exon 9 and *CAMK2δ*-NLS/exon 14 splicing were affected in LVNC-CMs exclusively, which was rescued in resLVNC-CMs (Fig. 21B). Increased exon 5 inclusion in *LDB3* was shown for DCM-CMs and consistently splicing is reinstated in resDCM-CMs (Fig. 21C). Of note, even though without statistical significance, *LDB3* splicing is also affected in LVNC-CMs (p=0.09). Furthermore, *RBM20* downregulation in LVNC-CMs is rescued in resLVNC-CMs (Fig. 21D) indicating that the RBM20 mutation p.R634L in LVNC has an influence on its own expression levels, whereas the p.R634W mutation in DCM does not.

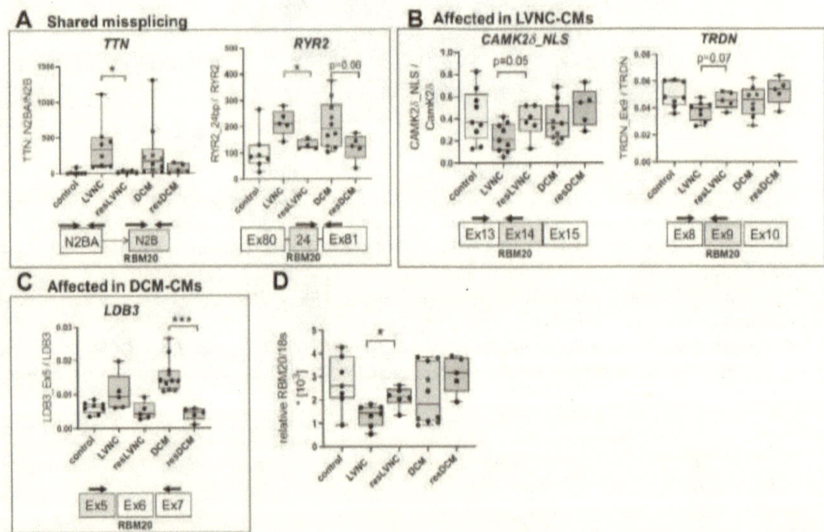

Fig. 21 RBM20 rescue-lines validated the RBM20-dependent missplicing in LVNC- and DCM-CMs
A-D: Data is represented as box plots, whereas every dot represents one cardiac differentiation. Statistical analysis is by Student's t-test of resLVNC vs LVNC and resDCM vs DCM and results are marked with * p<0.05, ** p<0.01 and *** p<0.001.
A: Shared missplicing of *TTN* and *RYR2* in LVNC- and DCM-CMs rescue-iPSC-CMs.
B: Missplicing of LDB3 in DCM-CMs validated by resDCM-CMs.
C: Missplicing of CAMK2d and TRDN in LVNC-CMs validated by resLVNC-CMs
D: RBM20 is downregulated in LVNC-CMs validated by resLVNC-CMs.

Next, the resLVNC and resDCM-CMs were assessed for structure, BR and Ca^{2+} handling to investigate the effects of the rescue of the RBM20 mutation. Sarcomeric regularity was analyzed as described previously. As depicted in Fig. 22A, resLVNC- and resDCM-CMs showed a regular sarcomeric pattern. Quantification of the Z-disc regularity underscores this observation as regularity values of resLVNC- and resDCM-CMs improved significantly and are comparable to the control-CMs (Fig. 22B). This data demonstrated that RBM20 mutations cause sarcomeric disarray, which is possibly connected to the missplicing of *TTN*. MEA measurements of resLVNC- and resDCM-CMs showed comparable values for BR, which is on average 0.6 Hz (Fig. 22C). Lastly, Ca^{2+} transients were recorded by using Fluo-4. ResLVNC-CMs showed significantly elevated levels of basal Ca^{2+} rise and decay time compared to LVNC-CMs and did not differ from values previously shown for control-CMs. In line with this observation, a prominent reaction to Iso stimulation was observed in resLVNC-CMs as opposed to the weak reaction previously observed in LVNC-CMs (Fig. 22D). Ca^{2+} leakage that was significantly increased in DCM-CMs showed significant reduction in the resDCM-CMs (Fig. 22E).

Taken together, this data showed that the distinct RBM20 mutation in LVNC and DCM is responsible for the shared disease phenotypes of sarcomeric disarray. Intriguingly, the differential disease phenotypes in Ca^{2+} handling were also ameliorated in the isogenic control lines. These findings underscore that mutations in RBM20 can lead to different disease outcomes even if the same aa position is affected by a synonymous mutation. Isogenic control lines are a powerful tool to attribute pathologies to a gene mutation identifying RBM20 p.634 as a crucial position for disease variants.

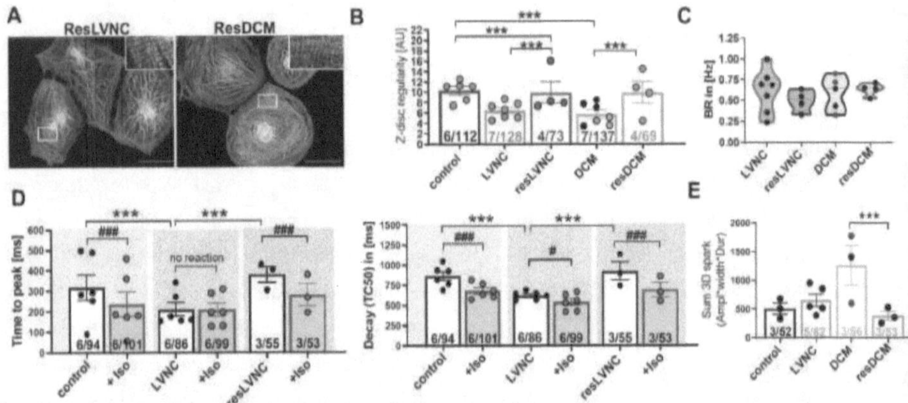

Fig. 22 Physiological impairments are abolished in isogenic RBM20-rescue lines of LVNC and DCM
A: Visualization of the sarcomere by immunofluorescence stainings against α-actinin (green) and titin M8/M9 (red) for resLVNC- and resDCM-CMs (scale bars = 50 μm).
B: Sarcomeric regularity is restored in resLVNC- and DCM-CMs. Data is presented as nested mean+/- SEM; p-values were calculated by nested one-way ANOVA multiple comparisons with Dunnett correction and results are marked with * p<0.05, ** p<0.01 and *** p<0.001. The numbers in the graph represent [No. of differentiations/analyzed pictures].
C: Beating rate (BR) measurements. Every dot represents one cardiac differentiation. Data is presented as violin plots and statistics were calculated by one-way ANOVA multiple comparisons with Dunnett correction, but no significances were detected.
D: Ca^{2+} cycling is restored in resLVNC-CMs. Transient rise **(left)** and decay **(right)** time is elevated in resLVNC-CMs and subsequently resLVNC-CMs showed a strong reaction to Iso. Data is presented as nested mean+/- SEM. The numbers in the graph represent [No. of differentiations/analyzed cells]. P-values were calculated by nested one-way ANOVA against control of the basal condition with Dunnett correction and results are marked with * p<0.05, ** p<0.01 and *** p<0.001. Additionally, #p-values were calculated for basal und Iso-treated condition for each group using nested t-test and results are marked with # p<0.05, ## p<0.01 and ###p<0.001.
E: Ca^{2+} leakage is decreased in resDCM-CMs. The numbers in the graph represent [No. of differentiations/analyzed cells].Data is presented as mean+- SEM; p-values were calculated by nested t-test DCM vs resDCM and results are marked with * p<0.05, ** p<0.01 and *** p<0.001.

4.5 NOVEL RBM20 SPLICE-TARGETS

4.5.1 Screening for novel RBM20 splice targets by next generation sequencing technology

With isogenic RBM20-CMs available for LVNC- and DCM-CMs, the question was addressed if novel RBM20 splice targets can be identified. For this, mRNA of 90 d old control-, LVNC-, DCM, resLVNC-, and resDCM-CMs was isolated and Illumina NGS was performed by the AG Meder at the university hospital in Heidelberg for transcriptome analysis. The data was analyzed using the DexSeq pipeline to assess the expression levels of exonic bins within a gene. To identify possible new splice targets, the following steps were carried out: **1**. LVNC-CMs and DCM-CMs were analyzed against control- and rescue-CMs to compare differential exonic bin expression (by AG Meder, Dr. Weng-Tein). All genes/exonic bins with a p_{adj} value >0.05 were excluded. This resulted in 9854 genes with differential exon expression for LVNC (vs control- and resLVNC-CMs) and 10819 genes for DCM (vs control- and resDCM-CMs). Of these, 8947 genes are shared between the LVNC and DCM cohort. (Fig. 23A) **2**. GOterm analysis was performed on the shared genes between LVNC and DCM cohort. The GOterm "Cardiomyopathy" was the biggest hit (Fig. 23B) and 36 candidate genes were selected including different K$^+$ channels (*KCNQ1, KCNH2, KCNE1, KCNA4, KCNQ3, KCN1, KCND3*), Obscurin (*OBSCN*), *TRDN, RYR2, CACNA1c, SCN5A, SERCA2a, LMNA, MCU*, histone deacetylase 9 (*HDAC9*), adenylate cyclase 5 and 6 (*ADCY5, ADCY6*), ATP-binding cassette, sub-family C member 9 (*ABCC9*), myosin-binding protein 3 (*MYBPC3*), Ser-Arg rich splicing factor 3 (*SRSF3*), splicing factor 3b subunit 1 (*SF3B1*), RALBP1 associated Eps domain containing 2 (*REPS2*), Neurogenic locus notch homolog protein 2 (*NOTCH2*), SMAD specific E3 ubiquitin protein ligase 1 (*SMURF1*), Nexilin (*NEXN*), bone morphogenic protein 4 and 7 (*BMP4, BMP7*), ADAM metalloprotease domain 10 and 17 (*ADAM10, ADAM17*), cullin 3 (*CUL3*), calmodulin 1 (*CALM1*), Myosin light chain kinase 1 and 3 (*MYLK, MYLK3*), *RBM24* and *RBM20* itself. **3:** The exonic bins within every gene were translated into the respective exons of a gene und primers were designed to span the region of interest. All genes were tested in semi-quantitative PCR for the expression of different splice forms indicated by two or more bands. If more than one band was detected, the gel bands were cut and sent for Sanger sequencing to verify, which exons were included in a transcript. Subsequently, primers for qPCR analysis were designed to target the differentially expressed exon (same process as described in Fig. 12A). Fig. 23C shows all investigated genes that did not show multiple bands/isoforms and were excluded. In total, 30 of 36 screened genes were negative at this step of investigation.

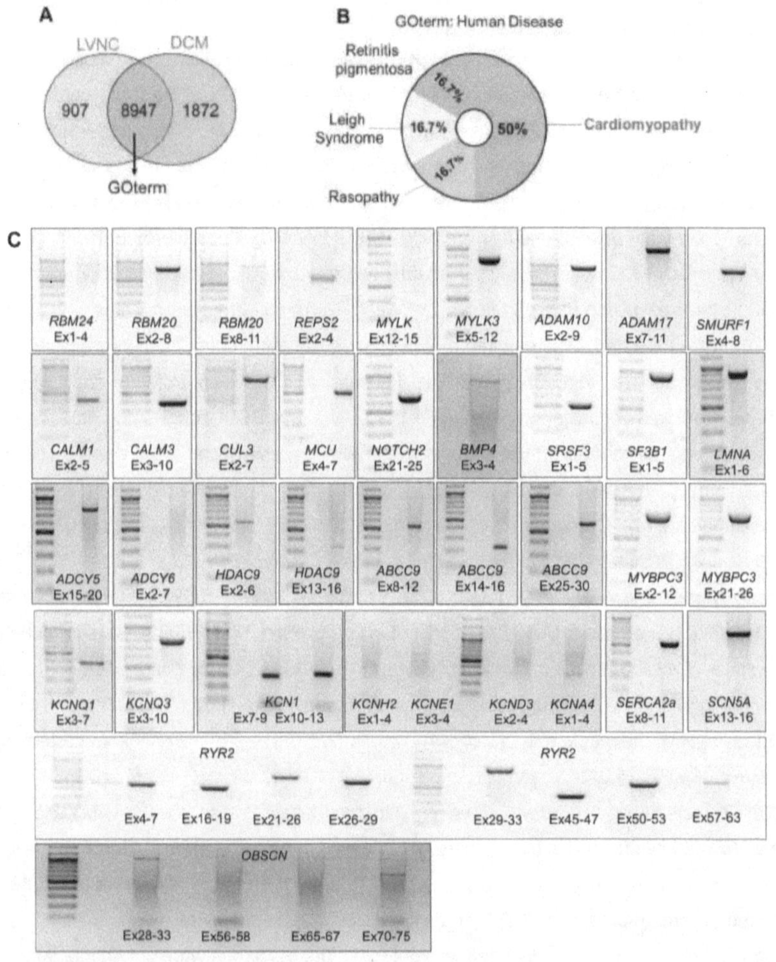

Fig. 23 NGS analysis
A: Venn diagram of genes with differential expressed exons.
B: GOterm analysis of the overlapping group reveals "Cardiomyopathy" as biggest hit. GOterm analysis for "Human Diseases" with a pvalue > 0.05.
C: Selected candidate genes that could not be verified.

4.5.2 Novel RBM20 splice targets *KCNH2, CACNA1C, ADCY6* and *BMP7*

From 36 screened candidate genes, six could be verified with different isoforms. Affected exons were verified by Sanger sequencing. The following genes were identified: exon 6 of *MYLK*, exon 6 of *KCNH2*, exon 13 of *CACNA1c*, exon 14 of *ADCY6*, exon 4 of *BMP7* and exon 10 of *NEXN* (Fig. 24A). In a next step, the levels of transcripts containing these exons of interest were quantified. For this experimental set-up, the isogenic lines of LVNC and DCM are indispensable to serve as control lines that reduce background from genomic variation among individuals. Two important ion channels, *CACNA1C* and *KCNH2,* were detected to be affected in LVNC- and DCM-CMs (Fig. 24B). Notably for *KCNH2* splicing, there were no significances detected between control-CMs and LVNC- or DCM-CMs. However, this target was counted as affected since there is a clear rescue effect observed in resLVNC- and resDCM-CMs. Furthermore, splice aberrations in *ADCY6* and *BMP7* were detected in DCM-CMs exclusively (Fig. 24C). Splicing defects of *MYLK* and *NEXN* was not verified for LVNC- or DCM-CMs (Fig. 24D). These results indicate that RBM20 splice targets are not fully elucidated yet. The DexSeq data contained many information and possible target genes.

In summary, the analysis showed four new possible splice targets. Most intriguingly, the important ion channels *KCNH2* and *CACNA1C*. *CACNA1C* was described before for RBM20-dependent splicing, however this analysis reveals that exon 13 is affected instead of exon 9, which was reported as RBM20-dependent in rat models (Guo et al., 2012). *KCNH2* codes for the hERG channel, an outward K^+ channel. This is the first report of RBM20-dependent splicing of a K^+ channel. Finally, *ADCY6* and *BMP7* are genes affiliated with cAMP cycling (ADCY6) and development (BMP7). These novel RBM20 targets are interesting since established RBM20 splice targets mostly comprise sarcomeric and ion channel genes. In conclusion, the using DexSeq was feasible to screen for differential exon expression within a gene. With this approach, *KCNH2, CACNA1C, ADCY6* and *BMP7* were identified and validated as new splice targets of RBM20.

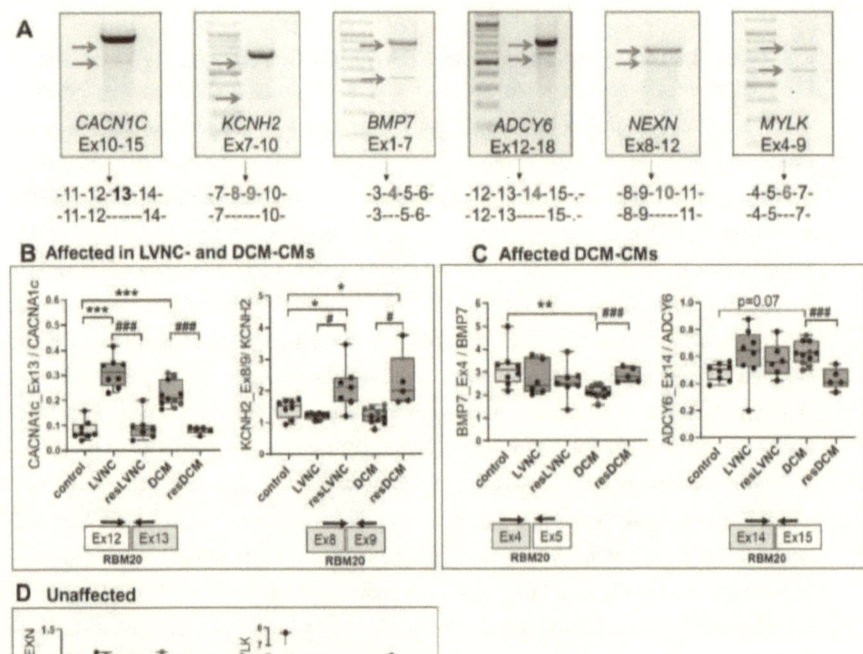

Fig. 24 Novel RBM20 splice targets
A: Sanger Sequencing of different PCR products revealed the RBM20-specific exon for splicing in target genes. Affected exons are marked in red.
B-D: QPCR results for the novel RBM20-target genes. Every dot represents one differentiation experiment. P-values are determined by one-way ANOVA against control with Dunnett correction and results are marked with * p<0.05, ** p<0.01 and *** p<0.001. Additionally, Student´s t-test was conducted for LVNC vs resLVNC and DCM vs resDCM and results are marked with # p<0.05, ## p<0.01 and ### p<0.001. Schematic under every graph depicts the RBM20-dependent exons for splicing (in yellow). Arrows indicate the positions were the primers bind.
B: Missplicing of *CACNA1C*-exon 13 and *KCNH2*-exon 8/9 is detected in LVNC- and DCM-CMs.
C: Missplicing of *BMP7*-exon4 and *ADCY6*-exon14 is detected in DCM-CMs exclusively.
D: Possible RBM20 splice targets *NEXN* and *MYLK* were not affected in LVNC- and DCM-CMs.

4.6 ABERRANT PHOPHORYLATION OF Ca²⁺ HANDLING PROTEINS IN LVNC-CMs

Fast Ca^{2+} kinetics and no prominent reaction to Iso are strong phenotypes for LVNC-CMs. The rate of Ca^{2+} cycling is predominantly regulated by phosphorylation of cardiac proteins such as the RYR2 and the SERCA2a accessory protein Phospholamban (PLN). Upon β-adrenergic stimulation, PKA and CAMK2δ are activated to phosphorylate RYR2 and PLN to increase Ca^{2+} cycling rate (Mattiazzi & Kranias, 2014). Intrigued by the data that *CAMK2δ* is a splice target exclusively affected in LVNC-CMs, the question was addressed if CAMK2δ-dependent phosphorylation is a driver for the fastened Ca^{2+} cycling in LVNC-CMs. For this, Western blot pellets were prepared, consisting of pairs of basal (unstimulated) and Iso treated samples to investigate the basal phosphorylation status of Ca^{2+} handling proteins RYR2 and PLN and their increase in phosphorylation after the addition of Iso. First, investigation of RYR2 and its PKA- and CAMK2δ-dependent phosphorylation sites (RYR2_Ser2808 and RYR2_Ser2814 respectively) were quantified. This analysis did not uncover any deviating phosphorylation status of RYR2 at basal level or after Iso stimulation in LVNC- and DCM-CMs compared to their respective isogenic controls (Fig. 25). Notably, there was no prominent phosphorylation increase observed for RYR2 after the addition of Iso (Fig. 25B and E).

In a next step, PKA- and CAMK2δ-dependent phosphorylation sites for PLN were quantified. For PLN, a reaction after Iso stimulation was observed, which is displayed by a prominent increase of phosphorylation (Fig. 26A and D). For DCM-CMs, basal levels of PLN phosphorylation for the PKA site PLN_Ser16 and the CAMK2δ site PLN_Thr17 showed comparable levels to resDCM-CMs (Fig. 26B and E) and a 3-4-fold increase in phosphorylation after Iso treatment as observed in resDCM-CMs (26C and F). In contrast, the CAMK2δ-dependent PLN_Thr17 site revealed elevated phosphorylation levels for LVNC-CMs at basal (Fig. 26B) and subsequently a significantly weaker phosphorylation increase after Iso stimulation (Fig. 26 C). This observation coincides with missplicing of *CAMK2δ* in LVNC-CMs, which was reversed in resLVNC-CMs. In comparison, LVNC-CMs exhibited comparable phosphorylation levels at basal and Iso stimulated conditions for the PKA-dependent PLN_Ser16 site (Fig. 26E and F). In conclusion, this data suggests that *CAMK2δ* missplicing leads to increased basal phosphorylation of PLN_Thr17 and weakened phosphorylation response after Iso that contributes to the fastened Ca^{2+} cycling rate and impaired Iso reaction observed in LVNC-CMs, which is ameliorated in resLVNC-CMs. In contrast, DCM-CMs do not share these pathologies in Ca^{2+} cycling and phosphorylation aberrations, which is congruent with the observation that CAMK2δ splicing is not affected. This identifies the RBM20 splice target *CAMK2δ* as s strong driver for disease phenotypes and splicing of *CAMK2δ* can determine different disease outcomes observed for different RBM20 variants.

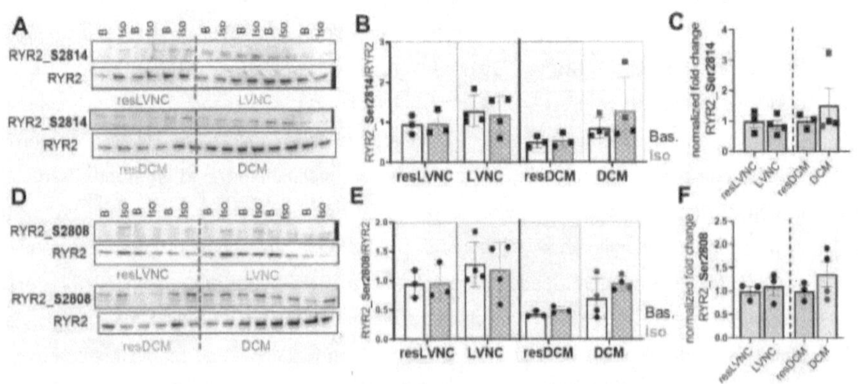

Fig. 25 Analysis of PKA- and CAMK2δ-dependent phosphorylation of RYR2
A: Western blot membranes for PLN and the CAMK2δ-site RYR2_Ser2814 at basal and Iso (1 µM) stimulation.
B: Quantification of the Western blot membranes for basal (black) and Iso-treated (green) samples for RYR2_Ser2814.
C: Normalized fold change (resLVNC and resDCM set as 1) in phosphorylation increase of RYR2_Ser2814 after Iso stimulation.
D: Western blot membranes for RYR2 and the PKA-site RYR2_Ser2808 at basal and Iso stimulation (1 µM).
E: Quantification of the Western blot membranes for basal (black) and Iso-treated (green) samples for RYR2_Ser2808.
F: Normalized fold change (resLVNC and resDCM set as 1) in phosphorylation increase of RYR2_Ser2808 after Iso stimulation.
B+C, E+F: Every dot represents one cardiac differentiation. P-values were calculated by Student's t-test and results are marked with * p<0.05, ** p<0.01 and *** p<0.001. No significances were detected.

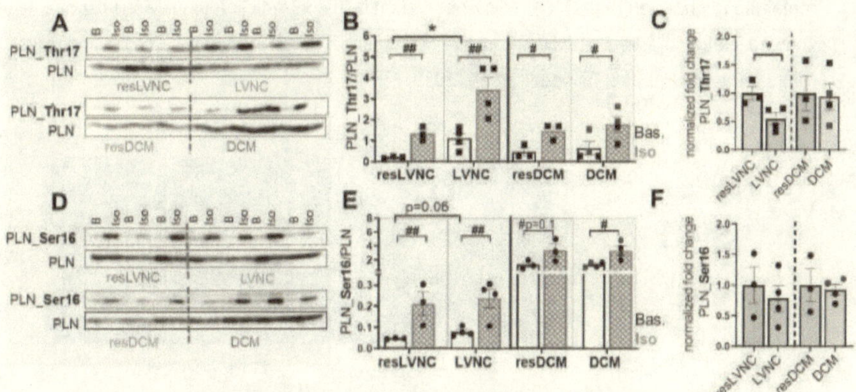

Fig. 26 Analysis of PKA- and CAMK2δ-dependent phosphorylation of PLN
A: Western blot membranes for PLN and the CAMK2δ-site PLN_Thr17 at basal and Iso stimulation (1 µM).
B: Quantification of the Western blot membranes for basal (black) and Iso-treated (green) samples for PLN_Thr17.
C: Normalized fold change (resLVNC and resDCM set as 1) in phosphorylation increase of PLN_Thr17 after Iso stimulation.
D: Western blot membranes for PLN and the PKA-site PLN_Ser16 at basal and Iso stimulation (1 µM).
E: Quantification of the Western blot membranes for basal (black) and Iso-treated (green) samples for PLN_Ser16.
F: Normalized fold change (resLVNC and resDCM set as 1) in phosphorylation increase of PLN_Ser16 after Iso stimulation.
B+C, E+F: Every dot represents one cardiac differentiation. P-values were calculated by Student's t-test and results are marked with * p<0.05, ** p<0.01 and *** p<0.001.

4.7 LOCALIZATION STUDIES

4.7.1 Localization of RBM20 in LVNC- and DCM-CMs

It was shown previously that the RS domain in RBM20 also functions as an NLS when phosphorylated (Murayama et al., 2018) (Fig. 27A). Since the RBM20 mutation in LVNC- and DCM-CMs affects the RS domain, it was tested if RBM20 distribution is altered. IF stainings detected RBM20 predominately in the nucleus with little detection in the cytoplasm (Fig. 27B). To test this further, the iPSC-CMs were separated into fractions of nucleus (Nuc)-, cytoplasm (Cy)- and chromatin-bound (Ch.b)-fraction to test the distribution of RBM20. The protein quantification showed significantly reduced amounts in the Cy-fraction and elevated levels in the Nuc-fraction for LVNC- and DCM-CMs. In contrast, RBM20 levels in the Ch.b-fraction remained unchanged (Fig. 27C). This observation is in contradiction to previous reports, which showed elevated levels of RBM20 in the cytoplasm if the RS domain is deleted or mutated (Murayama et al., 2018; Filippello et al., 2013).

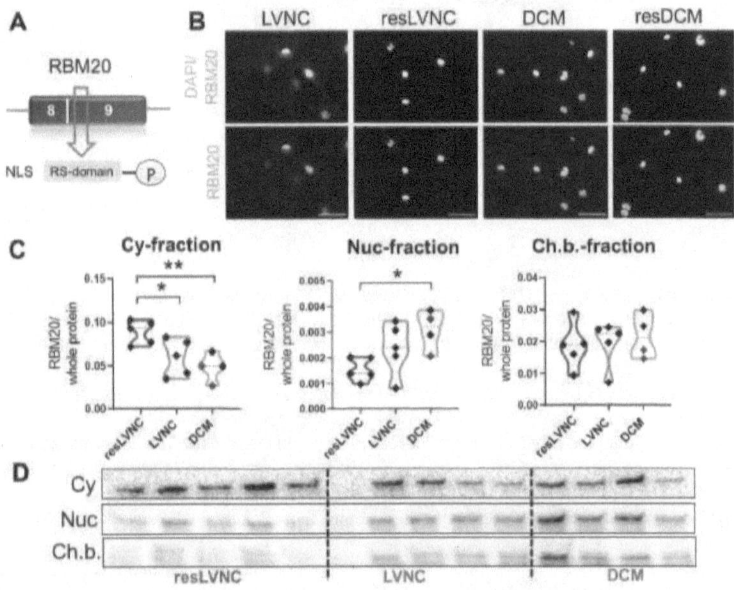

Fig. 27 Distribution RBM20 in LVNC- and DCM-CMs
A: Sheme of the nuclear localization sequence (NLS) described for RBM20.
B: Immunofluorescence stainings of RBM20 (yellow) in LVNC- and DCM-CMs. Scale bars = 50 µm
C: Protein quantification of RBM20 in the Cy, Nuc and Ch.b.- fraction. Every dot represents 1 cardiac differen-tiation. Statistical significance was tested with one-way ANOVA against resLVNC with Dunnett correction, and significant results are marked with * p<0.05, ** p<0.01 and *** p<0.001.
D: Original membranes with RBM20 stainings for the Cy, Nuc and Ch.b fractions.
Cy: cytsoplasm, Nuc: nucleus, Ch.b: chromatin-bound

4.7.2 Localization of CAMK2δ in LVNC- and DCM-CMs

LVNC-CMs showed accelerated Ca^{2+} kinetics and *CAMK2δ* missplicing, which is abolished in resLVNC-CMs. The RBM20-dependent exon 14 codes for an NLS in *CAMK2δ* (Fig. 28A), therefore the question was addressed, if CAMK2δ localization is altered in LVNC-CMs. IF stainings for CAMK2δ showed detection of CAMK2δ throughout the CMs (Fig. 28B). To test this further, the iPSC-CMs were separated into fractions of nucleus (Nuc)-, cytoplasm (Cy)- and membrane (Mem)-fraction to test the distribution of CAMK2δ. Quantification of CAMK2δ protein levels in the Cy-, Nuc- and Mem-fractions showed no significant difference compared to resLVNC-CMs (Fig. 28C), even though RBM20-dependent splicing concerns an NLS in *CAMK2δ*.

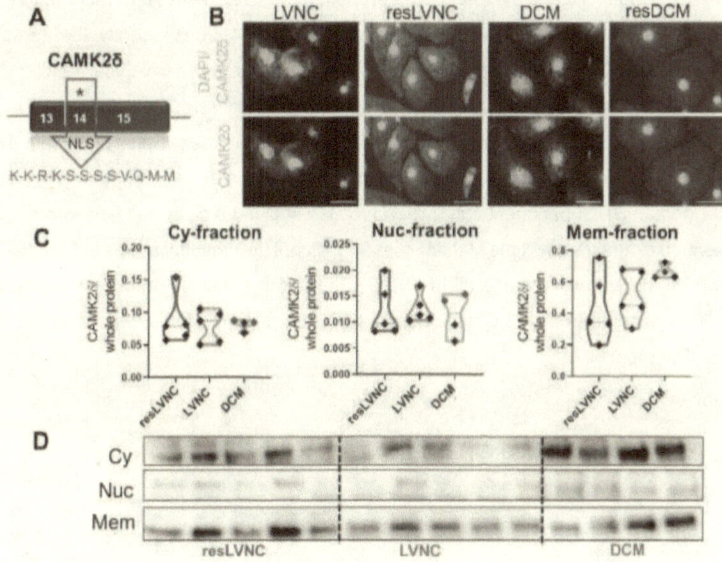

Fig. 28 Distribution RBM20 in LVNC- and DCM-CMs
A: Scheme of the nuclear localization sequence (NLS) described for CAMK2δ. * *CAMK2δ* missplicing was observed in LVNC-CMs.
B: Immunofluorescence stainings of CAMK2δ (cyan) in LVNC- and DCM-CMs. Scale bars = 50 µm
C: Protein quantification of CAMK2δ in the Cy, Nuc and Mem.-fraction. Every dot represents one cardiac differentiation. Statistical significance was tested with one-way ANOVA against resLVNC with Dunnett correction and significant results are marked with * p<0.05, ** p<0.01 and *** p<0.001.
D: Original membranes with CAMK2δ stainings for the Cy, Nuc and Mem fractions.
Cy: cytsoplasm, Nuc: nucleus, Mem: membrane

4.8 VERAPAMIL AMELIORATES THE Ca^{2+} PHENOTYPE IN LVNC-CMs

RBM20 p.R634L-based LVNC-CMs showed fastened Ca^{2+} cycling and impairment to β-adrenergic stimulation, which is conveyed by CAMK2δ hyperphosphorylation. ResLVNC-CMs showed improved Ca^{2+} homeostasis underscoring the RBM20 variant as a causative mutation. Reverting RBM20 mutations into wt would be a beneficial way for patients to lift their cardiac burden, however CRISPR/Cas9 mediated gene editing in humans is still at its infancy. Therefore, the question was addressed if the Ca^{2+} cycling phenotype in LVNC-CMs can be restored using pharmaceutical compounds. One promising candidate is Verapamil, an inhibitor of the L-type Ca^{2+} current with described negative chronotropic (reduction in beating frequency) effects (Krikler, 1986). To test the effects of Verapamil, Ca^{2+} transients were recorded at basal and after 15 min treatment with Verapamil (30 nM) using Fluo-4. Subsequently, the Verapamil-treated LVNC-CMs were stimulated with Iso (1 μM) to test if a physiological reaction to β-adrenergic stimulation can be restored. After the addition of Verapamil, a change in Ca^{2+} transients were observed, which presented as broadened Ca^{2+} transients (Fig. 29A). This is represented in increased transient rise and decay times, although the change was not significant (Fig. 29B). Subsequent stimulation with Iso showed a clear reaction with reduction in transient rise and decay time, which was significant for the decay time (Fig. 29C). In conclusion, treatment with Verapamil in LVNC-CMs had a beneficial effect on the fastened Ca^{2+} cycling observed in LVNC-CMs. Subsequently, the reaction to Iso was restored in the presence of Verapamil.

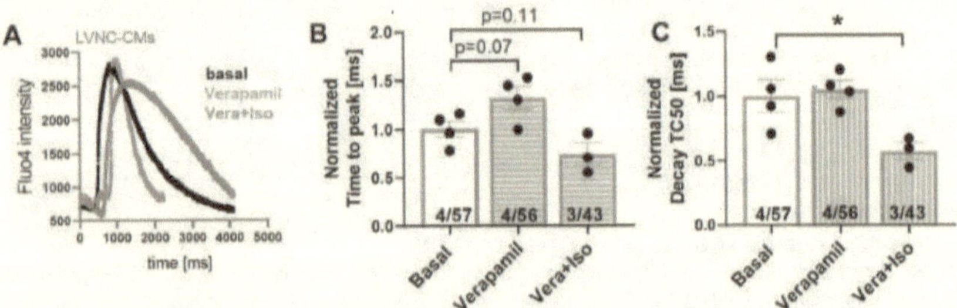

Fig. 29 Verapamil treatment in LVNC-CMs
A: Representative Ca^{2+} transient recordings at basal, Verapamil (30 nM) and Verapamil + Iso (1 μM) condition using Fluo-4.
B+C: Data is presented as nested mean+/- SEM; p-values were calculated by nested Student's t-test against the basal condition and results are marked with * p<0.05, ** p<0.01 and *** p<0.001. Multiple cells were analyzed [No. of differentiations/analyzed pictures] and are indicated in the graphs.
B: Ca^{2+} transient rise time and C: Ca^{2+} transient 50% decay time (TC50).

This identifies Ca^{2+} channel blockers like Verapamil as a prospective medical drug in the treatment of RBM20-based LVNC.

4.9 DEVELOPMENTAL ASSESSMENT

4.9.1 Developmental impairment in LVNC-CMs

LVNC is defined as a developmental disease with impaired myocardial compaction during cardiogenesis resulting in hypertrabeculated cardiac muscle in the adult heart (Arbustini et al., 2016). As mentioned above, a clinical study linked the RBM20-R634L mutation to LVNC (Sedaghat-Hamedani et al., 2017). Therefore, the question was raised if iPSC-CMs can be used to study developmental defects and if RBM20 has an influence in cardiac development. The cardiac differentiation protocol used involved cultivation for at least 60 d after cardiac induction. During these 60 d, the iPSC-CMs undergo maturation recapitulating CM development. Therefore, iPSC-CMs were analyzed at d30 to focus on an embryonic state. Although RBM20 is heart specific, it is not clear which cardiac and non-cardiac cells in particular express *RBM20* and are therefore affected by the mutation. In order to analyze *RBM20* expression in other cardiac cell types, iPSCs were differentiated into atrial iPSC-CMs and iPSC-ECs. For iPSC-ECs differentiation, CHIR99021 was added to the iPSCs to induce mesodermal commitment. Next, BMP4, VEGF and the TGF-ß inhibitor SB431542 were added to induce endothelial differentiation. At d7, the iPSC-ECs were purified by magnetic cell sorting using CD31 (Natividad-Diaz et al., 2019) (Fig. 30A). For atrial-CMs differentiation, the protocol is similar to the ventricular-CMs differentiation protocol, except for the addition of RA at d3-d6 (Cyganek et al., 2018) (Fig. 30B). The expression data underscores that RBM20 is a purely cardiac protein as only atrial- and ventricular-CMs showed *RBM20* expression, whereas skin fibroblasts and iPSC-ECs were negative (Fig. 30C). Next, it was checked if RBM20 is present during early time points of cardiac development. Quantification of mRNA-levels revealed that *RBM20* becomes rapidly upregulated during iPSC-CMs maturation. The iPSCs transcribe little amounts whereas at d30, *RBM20*-mRNA is significantly upregulated. Notably, *RBM20* mRNA-levels increased further during maturation except for LVNC-CMs (Fig. 30D). This shows that RBM20 is present during early cardiac development and can therefore be involved in early cardiogenesis. Next, cell growth was monitored as CMs should increase their cell size during maturation (Yang et al., 2014a). Control-CMs demonstrated a constant increase in cell size from d30 to d90. LVNC-CMs exhibited a retarded growth and were significantly smaller at d90.

Notably, DCM-CMs grew very big and had hypertrophic tendencies at d90 (Fig. 30E). Subsequently, the expression of developmental genes like Hairy/enhancer-of-split related with YRPW motif protein 2 (*HEY2)* and the myocyte enhancer-factor 2 B (*MEF2B*) (Xiang et al., 2006) mRNA-levels were assessed at d30 of differentiation for an initial screen to analyze whether expression levels of important developmental genes are affected in LVNC-CMs. *HEY2* and *MEF2B* were significantly upregulated in LVNC-CMs, compared to control-CMs. In contrast, DCM-CMs showed comparable levels to control-CMs (Fig. 30F). This is the first clue that cardiac development and maturation is affected in LVNC-CMs. Other studies on LVNC reported decreased proliferation in LVNC-CMs that affects cardiac compaction as a driver in LVNC development (Kodo et al., 2016). To assess whether the RBM20-based LVNC-CMs show a similar proliferation phenotype, d35 old iPSC-CMs were incubated for 48 hours with EdU to monitor DNA-synthesis (Fig. 30G). In addition to control-CMs, the isogenic lines were analyzed in order to minimize the bias of genomic heterogeneity. Indeed, the EdU-incorporation assay showed a lower proliferation rate in LVNC-CMs, which was elevated again in resLVNC-CMs (Fig. 30H). Of note, low attachment was observed for d30 old LVNC-CMs, which were difficult to digest as a monolayer. A smaller cell size in combination with decreased cell proliferation in LVNC-CMs could be a disease driver for impaired myocardial compaction observed in LVNC pathogenesis. Taken together, these data suggest a cardiac maturation phenotype in LVNC-CMs. This is the first evidence that RBM20 can be involved in cardiac development as LVNC-CMs, with the specific RBM20 variant p.R634L, were impaired in cardiogenesis.

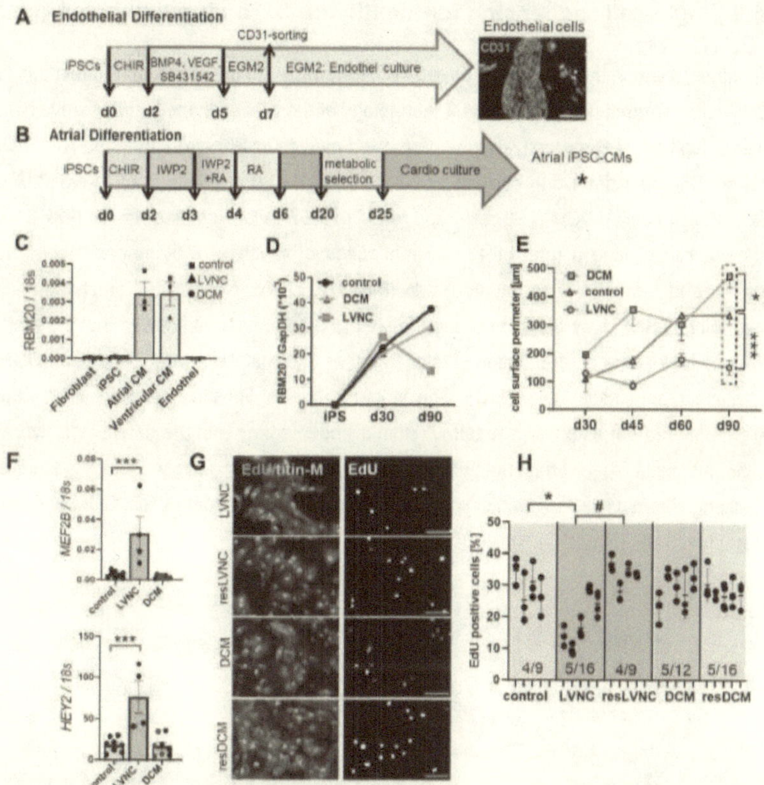

Fig. 30 Developmental impairments in LVNC-CMs
A: Differentiation protocol for iPSC-ECs.
B: Differentiation protocol for atrial iPSC-CMs. * Kindly provided by Wiebke Maurer
C: *RBM20* expression levels in different cell types.
D: *RBM20* expression levels in iPSCs and d30 and d90 old iPSC-CMs.
E: Cell growth (in cell surface perimeter) of iPSCs, d30 and d90 old iPSC-CMs.
F: Elevated expression levels of *HEY2* and *MEF2B* in LVNC-CMs. Every dot represents one differentiation experiment. P-values are determined by one-way ANOVA against control with Dunnett correction and results are marked with * $p < 0.05$, ** $p < 0.01$ and *** $p < 0.001$.
G: Immunofluorescence stainings of d35 old iPSC-CMs. Prior to staining the cells were incubated for 48 h with EdU (2.5 µM). Stainings against EdU (green) and titin-M (red). Scale bars = 50 µm
H: Quantification of EdU positive cells [%]. Data is presented as scatter plot. Numbers in the graph indicate [No. of differentiation/analyzed coverslips]; p-values were calculated by nested one-way ANOVA against control with Dunnett correction and results are marked with * $p < 0.05$, ** $p < 0.01$ and *** $p < 0.001$. Additionally, Student's t-test was conducted for LVNC vs resLVNC and DCM vs resDCM and results are marked with # $p < 0.05$, ## $p < 0.01$ and ### $p < 0.001$.

4.8.2 Single-cell sequencing for identifying CM maturation heterogenicity in LVNC-CMs

Preliminary data on LVNC-CMs maturation showed impaired development. To test this further, SCS was performed using the ICELL8 technology at the NGS-service for integrative genomics (NIG) in Göttingen. Data analysis was performed by Dr. Orr Shomroni from the NIG. SCS of a total of 1068 cells from one cardiac differentiation of 30 d old CMs from LVNC (293 cells), resLVNC (243 cells), DCM (265 cells) and resDCM (267 cells) were sequenced (Fig. 31A). The RNA-seq run yielded a total of 434.29 mio reads of which 97.5 % mapped to the human genome and furthermore to exonic reads (Fig. 31B). Next, t-distributed stochastic neighbor embedding (t-SNE) was performed to visualize the SCS data. This showed that the majority of cells from the same differentiation cluster together and that cells from LVNC and DCM form two distinct populations (Fig. 31C). The isogenic control lines cluster with their respective patient line and not as an own "healthy" cluster underscoring that the heterogenicity between individuals has a bigger effect than the disease phenotype. Intriguingly, there is a small subset of cells located outside the main clusters, which contains cells from LVNC, DCM and resDCM, but not resLVNC (Fig. 31C).

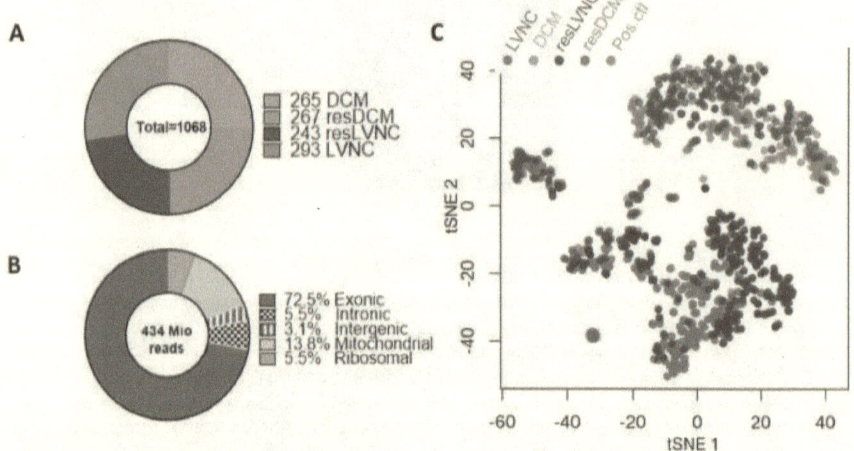

Fig. 31 Single-cell sequencing of iPSC-CMs at d30
A: Number of single cells, that were sequenced.
B: Distribution (%) of the 434 mio reads from SCS. The majority maps to exonic regions in the human genome with the remaining fractions mapping to intronic, intergenic, mitochondrial and ribosomal regions.
C: The tSNE plot of all sequenced cells from LVNC, resLVNC, DCM and resDCM.

Next, gene expression of selected genes of atrial markers *ISL1, NPPA, PTX2, HEY1, TNNI3*; ventricular markers *MYL2, HEY2, TNNT2*; cardiac transcription factors *NKX2.5, GATA4, GATA6, HAND1* and markers for maturation *RYR2, SCN5A, ATP2A2, MYH6, MYH7, TTN, CASQ2, CAMK2δ* and *RBM20* were analyzed for every cell. For better visualization the t-SNE plots were separated into 2 plots containing LVNC with resLVNC and DCM with resDCM to directly compare if gene expression is altered between the patient and the respective rescue line and if gene expression differs between patient lines. *GAPDH* expression was selected as a housekeeping gene and was consistently expressed among all conditions. The data indicates that the majority of cells from all lines show differentiation into cardiac cells highlighted by high expression of cardiac genes *RYR2, ATP2A2, TNNT2, CAMK2δ* and *RBM20* (data not shown). The expression of atrial markers *ISL1* and *PITX2* is increased in LVNC-CMs suggesting a tendency for atrial-like cells in LVNC-CMs. A marker for cardiac maturation is the ratio of *MYH6* to *MYH7* expression, where *MYH6* becomes downregulated in favor of *MYH7* expression during cardiogenesis. The SCS data showed comparable levels of *MYH7* expression, but increased levels of *MYH6 in* LVNC-CMs *(*Fig. 32*)*.

Taken together this analysis showed that the ICELL8 technology is a powerful tool for SCS in iPSC-CMs. Preliminary results indicate an impaired and delayed maturation phenotype in LVNC-CMs supported by higher expression levels of atrial genes *ISL1* and *PITX2* and the fetal *MYH6* compared to resLVNC- and DCM-CMs.

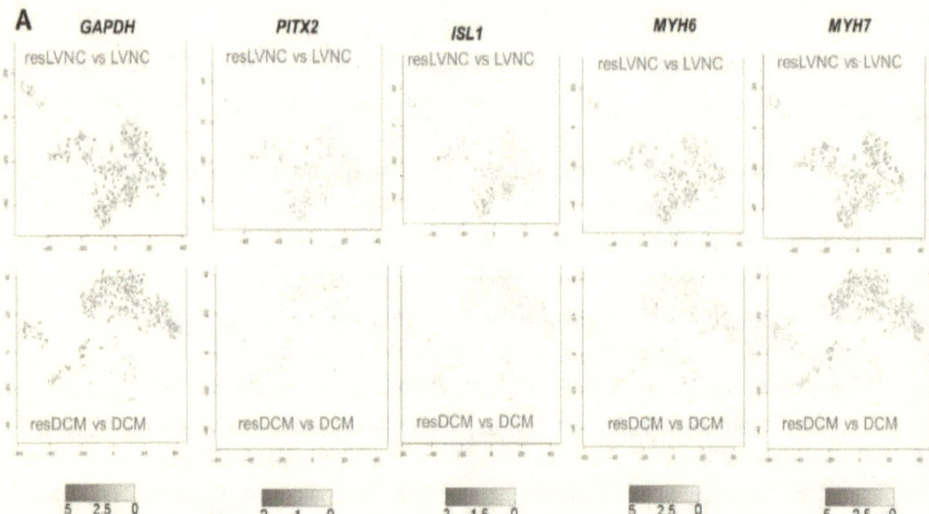

Fig. 32 Single-cell sequencing reveals upregulation of atrial and fetal markers in LVNC-CMs
Panels of cardiac gene expression patterns of *ISL1, PITX2, MYH6* and *MYH7*. *GAPDH* is shown as a housekeeping gene. Cells are colored based on the expression intensity of the respective gene. The atrial markers *ISL1* and *PITX2* and *MYH6* (marker for immaturity) is increased comparing the resDCM/DCM (upper) with the resLVNC/LVNC (upper panel). Furthermore, these genes are upregulated comparing resLVNC with LVNC within the (upper) panel.

5

DISCUSSION

In this project, RBM20-mutation based LVNC and DCM were investigated. RBM20 is an important cardiac splice factor that is essential for healthy cardiac function. To date, multiple genes have been identified to cause different cardiomyopathies, however little is known about the underlying pathomechanisms and how mutations in the same gene can cause different pathological outcomes. In this study, a patient-based model of RBM20-dependent LVNC and DCM was applied to study the underlying molecular and cellular pathomechanisms. Two families with hereditary heart conditions were identified to harbor two different missense mutations for the p.R634 in RBM20. LVNC patients of the one family contain a heterozygous missense mutation p.R634L, whereas the DCM patients of the 2nd family contain the heterozygous missense mutation p.R634W.

In the first part of this study, iPSC-lines were derived from one LVNC patient, two DCM patients and two healthy relatives from the DCM family. Later in this project, a second LVNC patient joined the study and her iPSC-CMs were used in the NGS experiments. For every patient, two iPSC-clones were cultured and characterized. The iPSC-lines showed expression of pluripotency markers and were capable to differentiate into cells from all three germlayers. Subsequently, the CMs derived from the iPSC-lines showed expression of cardiac markers, a sarcomeric structure and prominent beating in the culture dishes with high purity up to 96 %. The second part of this work compares the molecular and physiological outcomes of RBM20-dependent LVNC and DCM iPSC-CMs. Analysis of RBM20 splice targets showed shared missplicing in *TTN* and *RYR2*. Intriguingly, distinct missplicing was observed for *CAMK2δ* and *TRDN* in LVNC-CMs and for *LDB3* in DCM-CMs. The mRNA levels of *RBM20* were significantly reduced in LVNC-, but not in DCM-CMs. Subsequently, sarcomeric integrity, Ca^{2+} handling and electrical activity were investigated to assess if shared and differential phenotypes are observed. LVNC- and DCM-CMs showed an irregular sarcomeric structure and no differences in electrical activity compared to control-CMs. Ca^{2+} handling properties unveiled differential pathological manifestations. LVNC-CMs demonstrated accelerated Ca^{2+} kinetics, no prominent reaction to Iso stimulation and increased systolic Ca^{2+}. In contrast, DCM-CMs exhibited a prominent Ca^{2+} leakage, and decreased systolic Ca^{2+}. The only shared Ca^{2+} phenotype between LVNC- and DCM-CMs was the decreased diastolic Ca^{2+}. To assess if the disease phenotypes are linked to the RBM20 mutation exclusively, isogenic control-iPSCs were generated. For that purpose, the RBM20 mutations in LVNC and DCM were edited back into wt-RBM20 by CRISPR/Cas9 technology. The rescue-lines showed reversion of the pathological phenotypes in splicing, sarcomeric structure and Ca^{2+} handling, thereby confirming the RBM20 mutations as the disease-causing variants.

The third section of this book explored further pathomechanisms of the Ca^{2+} kinetics in LVNC-CMs, drug treatment with Verapamil, localization studies and novel RBM20 splice targets. *CAMK2δ* missplicing resulted in CAMK2δ-dependent hyperphosphorylation of targets such as PLN (Thr17) and was identified as a mechanism for the aberrant Ca^{2+} kinetics in LVNC-CMs, which was subsequently rescued in resLVNC-CMs. Furthermore, it was shown that the accelerated Ca^{2+} kinetics in LVNC-CMs can be ameliorated with the treatment of Verapamil. In contrast, aberrant localization of RBM20 or CAMK2δ was not observed in cell fractionation experiments. The last section of this book analyzed developmental parameters with a focus on LVNC-CMs. LVNC-CMs exhibited smaller cell size, increased expression of *MEF2B* and *HEY2* and showed decreased proliferation in 35d old iPSC-CMs. SCS supports the hypothesis of impaired development in LVNC-CMs demonstrated by increased levels of cardiac developmental atrial (*ISL-1, PITX2*) and fetal (*MYH7*) markers.

5.1 Somatic cell source–dependent reprogramming efficiencies

Somatic cells can be reprogrammed into pluripotent SCs by activating distinct transcriptional networks (Takahashi & Yamanaka, 2006). In this work, somatic cells from skin and blood samples were reprogrammed by ectopic expression of OKSM. In total, 15 iPSC-lines from LVNC (4 lines from 2 donors), DCM (4 lines from 2 donors), healthy DCM (4 lines from 2 donors) and unrelated healthy probands (3 lines from 2 donors) were generated and characterized for pluripotency. Since it has been shown that iPSC-lines harbor clonal differences (Ohnuki & Takahashi, 2015), two iPSC-lines were used for each patient. Skin fibroblasts were successfully reprogrammed with electroporation of OKSM coding plasmids with a reprogramming efficiency of 0.014 % or transduction with Sendai virus carrying OKSM with an efficiency of 0.01 %. Isolated PBMCs from blood samples were only reprogrammable by use of Sendai virus. Previous efforts by our laboratory to reprogram PBMCs with plasmids proved unsuccessful (Hübscher et al., 2019). In contrast to skin fibroblasts, PBMCs presented a greater challenge for reprogramming, highlighted by their low reprogramming efficiency of 0.002 % ± 0.001 (Hübscher et al., 2019). This is consistent with previous reports about blood reprogramming (Staerk et al., 2010). Firstly, only nucleated cells can be reprogrammed and therefore have to be isolated. In a blood sample, nucleated cells such as granulocytes, monocytes, lymphocytes and progenitor cells only comprise a small portion of blood cells. Secondly, the expansion of PBMCs is not as efficient as skin fibroblasts and requires multiple supplemental factors like EPO, SCF, IL-3, IL-6 and FLT-3. Lastly, PBMCs grow in suspension and need to adhere during the reprogramming process, whereas some of the forming iPSC-colonies were observed to detach 3 – 5 d later. The different cell sources and reprogramming methods in this study comprised fibroblasts reprogrammed with plasmids (LVNC 1), fibroblasts reprogrammed with Sendai virus (healthy DCM 1 and 2, DCM 2 and LVNC 2) and PBMCs reprogrammed with Sendai virus (DCM 1, Ctl 3, Ctl 4) (for details see Tab. 5). Although the cell sources and reprogramming method was not uniformly among all samples, the generated iPSC-lines did not show differences in pluripotency and cardiac/endothelial differentiation qualities. All iPSC-lines showed high activity of ALP, expression of pluripotency markers on mRNA (*OCT4, SOX2, NANOG, LIN28, GDF3*) and protein level (OCT4, SOX2, NANOG, LIN28, TRA1-60, SSEA4). Furthermore, all iPSC-lines were capable to differentiate into cells of the three germlayers, which was assessed by spontaneous *in vitro* differentiation.

To utilize the iPSC platform for cardiac research, the iPSCs were directly differentiated into CMs, for which previously published protocols were used (Burridge et al., 2012; Burridge & Zambidis, 2013; Lian et al., 2013). This protocol uses sequential addition of small molecules CHIR99021 and IWP2 in a chemically defined medium to induce cardiac differentiation. To enhance the CM population, a metabolic selection step was included around d 20 post differentiation start. This selection step uses glucose negative medium, which is supplemented with lactate as a carbon source. CMs are capable to metabolize lactate, whereas other cell types or cells with incomplete cardiac differentiation will be eliminated due to their dependency on glucose (Tohyama et al., 2013). The purity of iPSC-CMs was assessed with FLOW analysis after cTNT staining. This showed no differences among the different cell lines (91 % ± 6 cTNT positive cells), regardless of source material or reprogramming method. With regard to maturity, a long-term culture of >60 d was applied to enhance maturity. Our iPSC-CMs included polyploid cells (see IF images), expressed maturity marker like Nav1.5 or cTNT, showed Nav1.5-dependent ion currents and a clear sarcomeric structure as reported for prolonged culture duration (Kamakura et al., 2013; Lewandowski et al., 2018). Other features like T-tubuli or the capability for fatty-acid metabolism was not assessed. Maturity approaches were not a focus of this project and increased culturing of 60-90 d was a consistent procedure for all iPSC-CM experiments. However, it should be noted that the iPSC-CMs therefore resemble a fetal-like phenotype.

In summary, the cardiac differentiation and characterization was comparable among all 15 iPSC-lines regardless of the source material or reprogramming method. The cardiac differentiation protocol was suitable to produce a high number of CMs with high purity. Hence, this patient-specific iPSC platform was used to study RBM20 mutation-based LVNC and DCM in more detail.

5.2 Splice defects and RBM20 expression

RBM20 is a splice factor with essential splice targets in the heart. A study by Guo et al. identified approximately 30 genes using a knock out rat model (Guo et al., 2012). Based on these results, RBM20 splice targets were translated to the human locus in this work. With this, RBM20-dependent missplicing was identified for LVNC- and DCM-CMs as a molecular basis for disease development. Shared missplicing in LVNC and DCM-CMs included *TTN* and *RYR2*. *TRDN* and *CAMK2δ* were identified to be affected in LVNC exclusively, whereas *LDB3* missplicing was only observed in DCM-CMs. This suggests that differential missplicing is one driver for the development of different disease phenotypes.

5.2.1 Shared missplicing in LVNC- and DCM-CMs

RBM20 was the first splice factor identified to be involved in *TTN* splicing (Guo et al., 2012). TTN is one of the largest genes with 363 exons and a theoretical molecular weight of 4.2 MDa (if every exon is transcribed). TTN spans half a sarcomere from the Z-disc to the M-line providing structure and support for the thick filaments (Watanabe et al., 2018). During heart development and maturation, a TTN isoform switch occurs from the longer fetal TTN isoform N2BA to the adult short isoform N2B (Neagoe et al., 2002). It was shown that the fetal N2BA form increases in RBM20 knockout rats, underscoring the role of RBM20 in the exclusion of N2A to produce the adult TTN-N2B (Guo et al., 2012). TTN is a cardiomyopathy gene and altered ratios in N2BA to N2B have been observed in the failing heart. Furthermore, the fetal

N2BA form influences diastolic elasticity, leading to diastolic dysfunction when expressed in the adult heart (Neagoe et al., 2002). In line with these reports, altered N2BA/N2B ratios were observed in LVNC- as well as DCM-CMs in this work, whereas especially the lack of adult N2B contributed to the altered ratio (Fig. 13A). Both, RBM20 mutation p.R634W and L affected *TTN* splicing underscoring the importance of RBM20 as a major splice factor for *TTN*. Another study by our group analyzed the RBM20 mutation p.S635A in a patient-specific iPSC model of DCM. Similar to the results in this book, the S635A-iPSC-CMs showed altered N2BA/N2B mRNA expression. They showed further that N2B levels are also reduced on the protein level and linked the *TTN* missplicing to increased diastolic elasticity (Streckfuss-Bömeke et al., 2017). Based on this, it can be hypothesized that the LVNC- and DCM-CMs share this diastolic dysfunction due to the shared *TTN* splicing phenotype. To test this hypothesis, 3D models like engineered heart muscle (EHM) from LVNC- and DCM-CMs are necessary to investigate their mechanical properties. Preliminary results suggest that the diastolic elasticity is increased in LVNC- and DCM-CMs (data not shown). In addition, RBM20 was also implicated to generate circular RNA transcripts from the *TTN* gene, although the precise function of these transcripts remains to be determined (Khan et al., 2016).

RYR2 is a Ca^{2+} dependent Ca^{2+} release channel located in the SR. It is located in close proximity to voltage-dependent Ca^{2+} channels (Cav1.2) in the sarcolemma (Dulhunty, 2006). RBM20-dependent splicing of *RYR2* concerns a cryptic exon of 24bp length that arises from Intron 80-81 (Guo et al., 2012; Maatz et al., 2014). RBM20 is involved in removing this cryptic exon from *RYR2* transcripts, whereas in LVNC- and DCM-CMs, more transcripts with this 24bp-insertion were detected (Fig. 13A). Many *RYR2* mutations were linked to arrhythmia and tachycardia (George et al., 2006), whereas the *RYR2*-24bp variant in particular has been described to influence intracellular Ca^{2+} fluxes and have anti-apoptotic effects (George et al., 2007). Reports of RBM20 patients describe aggressive forms of DCM with high risk of arrhythmia (Brauch et al., 2009; Li et al., 2010; van den Hoogenhof et al., 2018) suggesting that the *RYR2*-24bp variant could be a contributing factor. To test this hypothesis, Ca^{2+} leakage and arrhythmic tendencies were assessed in LVNC- and DCM-CMs, which will be discussed below.

5.2.2 Missplicing exclusively in LVNC or DCM-CMs

CAMK2δ is an important player in cardiac Ca^{2+} homeostasis. This Ser/Thr kinase becomes activated by Ca^{2+}/calmodulin and influences cardiac Ca^{2+} kinetics by phosphorylation of its target proteins Cav1.2, RYR2, PLN and cTNI at specific positions. Aberrant functions of

CAMK2δ have been shown multiple times in cardiomyopathies and heart failure (Beckendorf et al., 2018). Alternative splicing produces multiple isoforms of CAMK2δ, whereas RBM20 is involved in inclusion of an NLS (Exon 14 in ENST00000508738.5), which is also defined as CAMK2δ$_B$ (Gray & Heller Brown, 2014). Splicing of Exon 14/NLS is strongly affected in LVNC-CMs of this study showing reduced levels of exon 14/NLS (Fig. 13B) and therefore reduced levels of the CAMK2d$_B$ isoform. Another CAMK2δ isoform is characterized by inclusion of exon 15/16 and defined as the CAMK2δ$_A$ isoform (Gray & Heller Brown, 2014). Reports about RBM20-dependent *CAMK2δ* missplicing in knockout rats and mouse ESC-CMs models also show downregulation of exon 14/CAMK2δ$_B$ with simultaneous upregulation of the exon 15/16/CAMK2δ$_A$ isoform (van den Hoogenhof et al., 2018; Guo et al., 2012; Beraldi et al., 2014). This isoform shift was not observed in LVNC- or DCM-iPSC-CMs of this study, as exon 15/16/CAMK2δ$_A$ expression was comparable to control-CMs. Notably, the splicing of exon 14 and exon 15/16 have been described as mutually exclusive in rats (Guo et al., 2012). However, the sequencing analysis of this book of *CAMK2δ* PCR products in control- and patient-CMs showed transcripts that harbor exon 14-15-16 (Fig. 12B), therefore contradicting them as mutually exclusive. This discrepancy could be attributed to inter-species differences and further investigation is needed to assess RBM20-dependent *CAMK2δ* splicing in a human model system like explanted human heart tissue. It has been proposed that altered RBM20-dependent *CAMK2δ* missplicing in knockout mice and human heart material has pathological effects on Ca^{2+} handling in CMs (van den Hoogenhof et al., 2018). Furthermore, the expression of CAMK2δ$_B$ was shown crucial in neonatal mouse cardiomyocytes in maturation and development of functional excitation-contraction coupling (Xu et al., 2005). This is reminiscent of RBM20-*TTN* splicing, where RBM20 is also involved in the production of the mature *TTN*-N2B isoform highlighting RBM20-deficiencies as driver to fetal-like phenotypes in cardiac diseases. Taken together, *CAMK2δ* missplicing was prominent in LVNC-, but not DCM-CMs, suggesting that LVNC-CMs have a Ca^{2+} handling phenotype, which is distinctive from the DCM-CMs.

TRDN is a SR-associated protein that links RYR2 with the regulatory Calsequestrin 2 (CASQ2) and Junctin protein and contributes to the regulation of Ca^{2+} release (Hancox et al., 2017). TRDN mutation carriers have been described with catecholaminergic polymorphic ventricular tachycardia underscoring the role of TRDN in intracellular Ca^{2+} release (Roux-Buisson et al., 2012). RBM20-dependent splicing of *TRDN* affects exon 9, which was downregulated in LVNC-CMs (Fig. 13B). Notably, the decreased levels of *TRDN*-exon9 is contradictory to increased *trdn*-exon9 levels in RBM20 knockout rats (Guo et al., 2012). This suggests that inter-species differences exist for *TRDN* or that RBM20 is overactive for this splice target. To conclusively determine whether RBM20 is responsible for *TRDN*-exon 9 inclusion or exclusion, RBM20 knockout studies in a human system are needed. Furthermore, the distinct role of exon 9 for *TRDN* transcripts remains elusive. However, the exclusive missplicing of *TRDN* and *CAMK2δ* in LVNC-CMs supports the hypothesis for a distinctive Ca^{2+} handling phenotype in LVNC-CMs, which will be discussed further in the following sections.

LDB3 is a protein that is localized at the sarcomeric Z-disc where it interacts with α-actinin and actin. LDB3 contains an N-terminal PDZ domain for interaction with kinases like protein kinase C and a C-terminal LIM domain for a-actinin binding. LDB3 functions for stabilization of the sarcomere (Krcmery et al., 2010). Furthermore, mutations in *LDB3* have been linked to the development of DCM and LVNC (McNally & Mestroni, 2017; Wang et al., 2019). This case of RBM20-dependent splicing, exon 5 in *LDB3* showed retention in DCM-CMs compared to control-CMs (Fig. 13B). As LDB3 gives structural support to the cardiac sarcomeres, it is suggested that, together with the *TTN* missplicing, this contributes in the sarcomeric disarray. Exon 5 of *LDB3* is not part of the PDZ or LIM domain and *LDB3* disease variants do not identify mutations located within exon 5. However, it has been proposed that exon 4 and 5 are mutually exclusive and that exon 4 is essential for binding of the glycolytic enzyme PGM1, whereas impaired binding to PGM1 contributes to the pathogenesis of DCM (Arimura et al., 2009). Notably, similar to *TRDN* splicing, *LDB3*-exon 5 splicing is conversely affected compared to the reports of RBM20 knockout rats (Guo et al., 2012). As discussed above, a human RBM20 knockout system is necessary to clarify this observation.

Other reported RBM20 splice targets such as *SORBS1*, *CACNA1C* and *CAMK2y* were not affected in LVNC- and DCM-CMs in this study (fig. 13D). This could be due to the fact that RBM20 is not involved in their splicing or has compensatory mechanisms in the human organism. Another possibility is that the respective mutations in LVNC and DCM do not have an effect on these targets. It should be noted that RBM20-dependent splicing of exon 9 in *CACNA1C* was not validated in this work. This target was identified in NGS experiments for RBM20 knockout rats, but the effect of altered exon 9 expression could not be validated by qPCR experiments (Guo et al., 2012). CACNA1C codes for the voltage-gated L-type Ca^{2+} channel Cav1.2, which initiates CICR. Studies in RBM20 null mice showed increased Ca^{2+} currents supporting the influence of RBM20 on Cav1.2 experimentally (van den Hoogenhof et al., 2018).

In summary, the analysis of RBM20 splice targets in LVNC- and DCM-CMs resulted in shared and differential missplicing. This suggests that common missplicing events will lead to shared disease manifestations while the distinctive missplicing of target genes will result in different disease phenotypes.

Expression levels of *RBM20* mRNA were significantly reduced in LVNC-CMs (Fig. 13E). This indicates that a haploid insufficiency mechanism is involved in the pathogenesis of LVNC-CMs. In support of this hypothesis, *TTN* and *CAMK2δ* missplicing was observed in RBM20 knockout mice and in mouse ESC-CMs after shRNA mediated RBM20 knockdown (Beraldi et al., 2014; van den Hoogenhof et al., 2018). Notably, DCM-CMs do not show *RBM20* decrease, but exhibit missplicing including *TTN* and *RYR2*, which shows that pathological effects can not be exclusively attributed to RBM20 downregulation. In this project, only *RBM20* mRNA levels were analyzed. To complement this data, the quantification of RBM20 protein amounts and a RBM20 degradation analysis is necessary to assess whether the mRNA downregulation in LVNC-CMs translates to the protein level.

5.3 Sarcomeric organization

Sarcomeres are the smallest contractile unit of a myocyte and the regular and repetitive arrangement of sarcomeres are essential for efficient cardiac contraction (Clark et al., 2002). LVNC- and DCM-CMs show a significant 1.6-fold reduction in sarcomeric Z-disc regularity compared to control-CMs, identifying a shared phenotype in LVNC and DCM (Fig. 14). Previous studies on RBM20 mutations also showed disturbed sarcomeric organization in S635A-iPSC-CMs (Streckfuss-Bömeke et al., 2017), R636S-iPSC-CMs (Wyles et al., 2016) and heart biopsis with P638L (Beraldi et al., 2014). The influence of RBM20 in *TTN* splicing is considered a main contributor for sarcomeric disarray, however, whether the impaired *TTN* isoform shift is the direct contributor for the sarcomeric disarray, or a secondary effect remains to be elucidated. In addition to sarcomeric disarray, altered expression of *TTN* N2BA/N2B isoforms is associated to influence diastolic function and passive stiffness (Nagueh et al., 2004; Warren et al., 2004). A 3D EHM model of RBM20-S635A-iPSC-CMs was able to assess mechanical characteristics. The RBM20-S635A-EHMs showed reduced force generation and a reduction in passive tension, which impacts diastolic function and filling (Streckfuss-Bömeke et al., 2017). Notably, the focus of many studies has been on the relationship of RBM20 and *TTN* splicing, whereas the influence of splicing on other sarcomere proteins like *LDB3* is less well studied. This data suggests that structural genes are major splice targets of RBM20 and that mutations, especially in the RS domain of RBM20, will cause sarcomeric irregularity.

Whether force generation and passive stiffness is impaired in LVNC- and DCM-CMs can be hypothesized from the irregular sarcomere structure and *TTN* missplicing, however further studies in single-cell myofilament assays and 3D EHMs are needed to test this hypothesis.

In this work, additional DCM iPSC-CMs with a genetic predisposition in splice factors and substrates were available at the lab and tested for sarcomeric integrity. DCM-iPSC-CMs with an underlying SCN5A/RBFOX1 and an LMNA mutation also showed significant sarcomeric disruption (1.4-fold ± 0.1 decrease in regularity compared to control-CMs). The AAV6 mediated knockdown of SLM2, a novel identified and so far, unpublished splice factor in the heart (cooperation with B. Meder and J.-N. Boeckel, Heidelberg), showed similar results with a 1.5-fold decrease in sarcomeric regularity. Further studies regarding myofilament function on a single-cell-level would give more insights into the functional consequences of the dysregulated sarcomeric structure.

Taken together, these findings add to the accumulating evidence that sarcomeric disarray is a major pathological driver in DCM. Due to this uniformly feature in all tested cardiac patient-lines, sarcomeric regularity assays could present a suitable read-out platform for drug screening experiments.

5.4 Ca^{2+} CYCLING in LVNC- and DCM-CMs

Ca^{2+} cycling is a crucial and tightly orchestrated mechanism in CMs. Ca^{2+} serves as a second messenger and actively converts the electrical stimulus into mechanical contraction of the ventricular CMs (Eisner et al., 1998). Here, Ca^{2+} kinetics and homeostasis were assessed in LVNC- and DCM-iPSC-CMs. First, Ca^{2+} cycling rise and decay time was measured using the Ca^{2+}-probe Fluo-4. The analysis revealed a significant decrease in rise and decay time for LVNC- and DCM-CMs compared to control-CMs. Strikingly, LVNC-CMs demonstrated a more prominent decrease, meaning they have a significantly faster Ca^{2+} cycling even compared to DCM-CMs (Fig. 16B). To test this further, the iPSC-CMs were stimulated with 1 µM Iso to examine if the cells can show a reaction to β-adrenergic stimulation. Indeed, control- and DCM-CMs show a physiological reaction to Iso, demonstrated by their significant decrease in rise and decay time. In contrast, LVNC-CMs show a weak reaction upon stimulation with Iso, possibly due to their already fast Ca^{2+} cycling at basal level (Fig. 16B). Ca^{2+} kinetics has been investigated in other RBM20 studies. Contrary, in iPSC-CMs models of RBM20-S634A and RBM20-R636S Ca^{2+} transient recordings showed increased time-to-peak values (synonym for rise time) and characteristics of Ca^{2+} extrusion (synonym to decay time) (Streckfuss-Bömeke et al., 2017; Wyles et al., 2016). However, comparison among these data sets proves to be complicated since the Ca^{2+} transients were recorded from spontaneous beatings in the mentioned studies, whereas in this study here, the cells were paced to obtain measurements, which are independent of frequency effects. Furthermore, the R636S-iPSC-CMs were measured at d24 (Wyles et al., 2016), which presents an earlier fetal time point compared with this study (measurements at d70 – d90). In addition to Ca^{2+} cycling, staining with the Fluo-4 probe allows to quantify Ca^{2+} sparks during the diastole. Ca^{2+} sparks are spontaneous releases from the RYR2, whereas increased Ca^{2+} sparks/leakage can facilitate arrhythmogenic events. Intriguingly, only DCM-CMs, but not LVNC-CMs, showed increased Ca^{2+} leakage. The tendency for increased Ca^{2+} leakage has been reported as well for RBM20 null mice and is suggested to contribute to the arrhythmia events, which frequently occur in RBM20 patients (van den Hoogenhof et al., 2018). Since Ca^{2+} leakage is a reoccurring observation in RBM20 models, it is surprising that LVNC-CMs do not show this phenotype suggesting that not every RBM20 mutation will lead to Ca^{2+} leakage or have another mode-of-action for the development of arrhythmic events.

To complement the Ca^{2+} analysis, the cells were stained with the ratiometric Ca^{2+}-probe FURA-2, which allows to assess absolute Ca^{2+} concentrations/levels. LVNC- and DCM-CMs showed reduced levels of diastolic Ca^{2+}. DCM-CMs also exhibited reduced systolic Ca^{2+}, whereas LVNC-CMs showed elevated levels of systolic Ca^{2+} (Fig. 17). In the physiological context, diastolic Ca^{2+} represent the phase of the relaxed myocardium to allow ventricular filling. Systolic Ca^{2+} represents the rapid increase of cytoplasmic Ca^{2+} to convey contraction, whereas the amount of Ca^{2+} positively correlates with force generation. Notably, decreased systolic Ca^{2+} in combination with increased diastolic Ca^{2+} is a reoccurring observation in DCM and heart failure (Eisner et al., 2020; Sen et al., 2000; Lou et al., 2012). Reports from other RBM20 studies showed similar results to this study, but should be interpreted with caution due to different model organisms and/or different RBM20 mutations. For RBM20 null mice, increased diastolic Ca^{2+} and systolic peak transients were reported (van den Hoogenhof et al., 2018). Measurements in S635A-iPSC-CMs showed decreased diastolic Ca^{2+} and increased systolic Ca^{2+} (Streckfuss-Bömeke et al., 2017), which mirrors the phenotype of the LVNC-CMs. In addition, increased systolic Ca^{2+} was also reported for R636S-iPSC-CMs and RBM20 knockdown in mouse ESC-CMs (Wyles et al., 2016; Beraldi et al., 2014). In contrast, the combination of decreased diastolic and systolic Ca^{2+} levels observed in DCM-CMs have not been reported yet. As Ca^{2+} cycling is linked to contractility properties in the heart (Vikhorev & Vikhoreva, 2018), this data indicates distinct diastolic and systolic dysfunctions for LVNC- and

DCM-CMs, which need further investigation of mechanical parameters. The different Ca^{2+} handling phenotypes are in line with the observation that Ca^{2+} related splice targets are differentially impacted in LVNC- and DCM-CMs. *TRDN* and *CAMK2δ* missplicing exclusively occurred in LVNC-CMs and could contribute to the fastened Ca^{2+} cycling observed. Notably, *LDB3* was the only differential splice target identified for DCM. Whether the splicing of a sarcomeric protein is a main driver for the differential Ca^{2+} properties observed in DCM remains elusive. The increased Ca^{2+} leakage in DCM-CMs indicates a dysfunctional RYR2. However, as *RYR2*-24bp transcripts were elevated for LVNC- and DCM-CMs, this missplicing can't be attributed solely to this phenotype.

In summary, LVNC-CMs show a strong pathological phenotype with fastened Ca^{2+} cycling and an impaired reaction to Iso stimulation. In contrast, DCM-CMs show a physiological reaction to Iso, but an increased Ca^{2+} leakage. Both, LVNC- and DCM-CMs showed reduced diastolic Ca^{2+} contents, but differential systolic Ca^{2+} amounts. This is the first report that demonstrated that two distinct mutations that concern the same aa position within a gene/RBM20 can lead to different pathological outcomes.

5.4.2 CAMK2δ-dependent hyperphosphorylation in LVNC-CMs

CAMK2δ missplicing was observed in LVNC-CMs, which was accompanied by accelerated Ca^{2+} kinetics and impaired reaction to Iso. Therefore, the influence of CAMK2δ was assessed by quantifying CAMK2δ-dependent phosphorylation at basal and Iso-stimulated state. LVNC-CMs showed significant basal increase in CAMK2δ-dependent phosphorylation of PLN, which targets the PLN_Thr17 position (Fig. 26B). Subsequently, a rapid increase in phosphorylation after Iso treatment was observed in all iPSC-CMs. However, the overall phosphorylation increase for LVNC-CMs was significantly lower (Fig. 26C). As phosphorylation is considered to increase Ca^{2+} cycling (Mattiazzi & Kranias, 2014), this observation correlates with the Ca^{2+} kinetics phenotype in LVNC-CMs. Firstly, the accelerated Ca^{2+} transient rise and decay times parallels the elevated PLN_Thr17 phosphorylation levels at basal state. Secondly, the impaired reaction to Iso in Ca^{2+} cycling relates to the weaker increase in PLN_Thr17 phosphorylation levels after Iso stimulation. In parallel, the phosphorylation of PKA was tested since this kinase shares phosphorylation targets with CAMK2δ and is also a major effector kinase after β-adrenergic stimulation. Basal and Iso-treated LVNC-CMs were comparable for their PKA-dependent PLN-Ser16 phosphorylation levels, which underscores CAMK2δ as the main contributor for aberrant phosphorylation in LVNC-CMs. In contrast, DCM-CMs, which did not show CAMK2δ missplicing, also did not show deviating phosphorylation at basal level and under Iso treatment highlighting the prominent RBM20-dependent *CAMK2δ* splicing in LVNC-CMs as a main contributor for their differential disease phenotype. Notably, phosphorylation of a second CAMK2δ-dependent target, RYR2, did not display any differences in phosphorylation levels at basal level or after Iso stimulation (Fig. 25). One limitation in this analysis was the difficulty to stain for RYR2 phosphorylation sites with antibodies, which resulted in weak signals and multiple bands on the Western blot membranes. To test whether CAMK2δ activity is generally increased or specific for the PLN target, additional CAMK2δ-dependent phosphorylation targets should be analyzed. This comprises CAV1.2 and cTNI, which should be quantified for their CAMK2δ- and PKA-dependent phosphorylation levels.

The influence of CAMK2δ in cardiac diseases has been well studied. Analysis of cardiac tissue in failing hearts revealed a 3-fold increase in CAMK2δ activity (Kirchhefer, 1999) and CAMK2δ$_B$ depleted transgenic mice show hypertrophy, ventricle dilation and decreased cardiac function (Zhang et al., 2002). Physiologically, a constitutive active CAMK2δ is associated with cardiac dysfunction and electrical instability (Swaminathan et al., 2012). Therefore, subsequent experiments for this study should include Westen blot and Ca^{2+} kinetics analysis of LVNC-CMs after treatment with a CAMK2δ-specific inhibitor.

Taken together, the *CAMK2δ* missplicing is a main contributor in LVNC-CMs with RBM20-R634L as a predisposition. In contrast, the DCM causing RBM20-R634W mutation does not result in CAMK2δ-missplicing or CAMK2δ-dependent hyperphosphorylation. This identifies CAMK2δ as a main disease-causing splice target for RBM20, whereas not all RBM20 mutations, like the DCM-causing p.R634W, will result in this disease phenotype. Furthermore, this is the first report of aberrant Ca^{2+} handling and CAMK2δ function in LVNC.

5.5 ELECTRICAL ACTIVITY

Electrical activity of LVNC- and DCM-CMs was assessed using the Multi Channel-MEA System, which allows to monitor BR, FPD and arrhythmic events. The Ca^{2+} data suggested possible phenotypes like increased BR due to fastened Ca^{2+} cycling and arrhythmia due to the increased Ca^{2+} leakage in DCM-CMs. However, no significant differences were detected in LVNC- and DCM-CMs compared to control cells with these MEA experiments. All iPSC-CMs showed a comparable BR of 0.64 Hz, 0.61 Hz, 0.58 Hz and a cFPD of 228 ms, 283 ms and 248 ms (control:LVNC:DCM). Stimulation with Iso showed an average 2-fold increase in BR in all tested lines. Arrhythmogenic tendencies were quantified by assessing the regularity of the interspike interval (ISI), which did not show a significant difference to the control-lines (Fig. 18).

However, patients with RBM20 mutations often suffer from arrhythmia and tachycardia (van den Hoogenhof et al., 2018; Watanabe et al., 2018; Hey et al., 2019), and CAMK2δ hyperactivity (as observed in LVNC-CMs) and increased Ca^{2+} leakage (as observed in DCM-CMs) is described to enhance arrhythmogenic tendencies (Swaminathan et al., 2012; Chelu & Wehrens, 2007). Therefore, further investigations of RBM20-dependent electrical activity with more sensitive methods should be performed in the future. In this MEA approach, a confluent monolayer of cells was digested onto electrodes, which measured the field potential in the extracellular space of the cells. Because of this, the measurements were influenced by the cell number and the proximity of the cells to the electrodes, which contributes to differences between measurements. An alternative approach could include single-cell recordings for the AP and/or single ion-currents, especially for the Cav1.2 channel.

5.6 RBM20 GENE EDITING

5.6.1 Gene editing of iPSCs with CRISPR/Cas9 technology

To study genetic-based diseases with the iPSC platform, choosing suitable control groups is crucial. For this, unrelated healthy donors and healthy family members serve as an invaluable tool. In this project, unrelated and related family donors were available. A limiting factor is the availability and recruitment of healthy family members. In case of the LVNC cohort no healthy relatives were available as a donor. Another factor to consider is the genetic background variability among individuals that could contribute to variable experimental outcomes. To conclusively study the contribution of a gene mutation, isogenic control lines are indispensable. In this work, one iPSC-line from LVNC and one iPSC-line of DCM were randomly selected for gene editing into wt-RBM20 using the CRISPR/Cas9 technology. Gene editing with CRISPR/Cas9 has been shown in many studies to efficiently create nucleotide substitutions, deletions and insertions in many model organisms (Pickar-Oliver & Gersbach, 2019). To edit one nucleotide in the genome, a distinct design of the guiding crRNA and HDR template is crucial for a successful edit. For the design of the crRNA, two major aspects were considered. Firstly, the crRNA needs to guide the Cas9 nuclease to the mutation and introduce the DSB in close proximity to the nucleotide mutation in accordance to existing PAM sites. Secondly, the crRNA needs to be evaluated with bioinformatic tools to determine off-target binding. Here, the Custom Alt-R CRISPR-Cas9 Guide RNA design tool from IDT was employed. This software analyzes the respective crRNA sequence by replacing every nucleotide with the other three possible nucleotides and predicting if this sequence is complementary to another position in the genome. The crRNAs chosen in this project have 3-4 mismatches before binding to off-target sites, whereas the predicted off-target sites do not locate into protein coding regions. The HDR template was designed as a ssDNA with 120 bp length, whereas the desired SNP editing sequence is located in the middle. To increase the CRISPR efficiency, silent mutations were introduced in the crRNA binding sequence or PAM site to prevent re-cutting events after a successful edit (Paquet et al., 2016). Furthermore, the crRNA and Cas9 were electroporated as a ribonucleoprotein since this has been proven more efficient, specific and less toxic than the use of plasmids (Liang et al., 2015; Zuris et al., 2015). In this project, two different crRNA/HDR template combinations were designed for LVNC and DCM, because the LVNC-causing RBM20 mutation destroyed a potential PAM site, which is present in wt and the DCM-causing variant (see Fig. 7 for CRISPR/Cas9 design). This approach was successful to generate isogenic control lines with wt-RBM20 with an editing efficiency of 45 % for LVNC and 2.8 % for DCM. This discrepancy underscores the high variability of genome editing in iPSCs reported in other studies, which range from 1 – 42 % (Byrne et al., 2014; Paquet et al., 2016; Xu et al., 2018). Although, the gene editing in this study concerned neighboring nucleotide positions, the CRISPR/Cas9 design for LVNC yielded a 16-fold higher editing efficiency. It was shown that the guiding crRNA has a substantial influence on the editing outcome and often the efficiency has to be tested experimentally (Xu et al., 2015; Zheng et al., 2017). This observation is supported by a study that obtained similar gene editing efficiencies when the same crRNA was used for different cell types (Xu et al., 2018). In this case, the crRNA used for LVNC-CMs proved more potent for the gene editing. Another observation was that some wt RBM20 edited lines showed the incorporation of the artificial missense mutations provided by the HDR-template, whereas other rescue-lines only harbored the edited wt RBM20 sequence (Fig. 19B). This suggests, that for DSB repair the other endogenous allele was used as a template instead of the exogenous provided HDR-template.

In conclusion, the generation of isogenic rescue lines for LVNC and DCM was successful. In total, 6 rescue lines were chosen for LVNC (4 resLVNC-iPSCs) and DCM (2 resDCM-iPSCs) and characterized. The selected rescue iPSC-lines showed all pluripotency and differentiation qualities observed in their respective counterparts. All rescue-lines showed expression of pluripotency markers on the mRNA and protein levels and were capable to differentiate into cells from the three germlayers. The rescue-lines were screened by Sanger sequencing to verify the successful edit into wt RBM20. However, subsequent sequencing of the RBM20 locus in later passages revealed weak signals for the initial RBM20 mutation. This was observed after 30 passages in one resLVNC-line (clone No 21) and one resDCM-line (clone No 35) (data not shown). This suggests, that the singularization of the iPSCs after the electroporation with CRISPR/Cas9 was not complete and mixed colonies were obtained. In this case, earlier passages of the respective rescue-clones were thawed and singularized again. This underscores the importance for frequent quality control to ensure the use of correctly edited iPSC-lines. Therefore, the rescue-iPSCs were sequenced every five passages to validate the use of edited iPSC-lines for cardiac differentiation. Another concern of the CRISPR/Cas9 system is the introduction of undesired off-target mutations (Li et al., 2019a), although reported off-targets by Cas9 in stem cells by whole genome coverage are considered rare (Veres et al., 2014). However, to fully validate off-target mutagenesis, whole genome sequencing of resLVNC and resDCM would be necessary with a special focus on the possible off-target sites predicted by the bioinformatic analysis. This would also unravel, whether the two CRISPR-designs that differ significantly in editing efficiency also show different rates in off-target editing.

In summary, the generation of isogenic rescue-lines of LVNC and DCM was successful. All rescue-lines showed wt-RBM20 sequence of p.R634R and maintained full pluripotency and differentiation potential characteristics.

5.6.2 Isogenic rescue lines to study RBM20 disease variants

The rescue-lines of LVNC and DCM were used to analyze the role of the RBM20 mutation in the molecular and cellular pathologies observed. Since only the RBM20 mutation was edited and the genetic background of the patients remained identical, the isogenic controls should elucidate the direct influence of the RBM20 variants. Cardiac differentiation was successful for all rescue-lines and showed comparable iPSC-CMs purities of 89 % ± 8 cTNT positive cells. Sarcomeric organization improved significantly in isogenic RBM20-lines, which is accompanied by the rescue of splicing in *TTN* and *LDB3*. Furthermore, the differential Ca^{2+} handling impairments were reverted in the respective rescue-lines. ResLVNC-CMs showed increased Ca^{2+} cycling rise and decay times with a strong reaction to Iso stimulation compared to LVNC-CMs. In line with this observation, the splicing of *RYR2*, *TRDN* and *CAMK2δ* was also reversed in resLVNC-CMs. Furthermore, the observed downregulation of RBM20 mRNA levels was elevated in resLVNC-CMs. As discussed earlier, studies on the protein level are necessary to elucidate the effect of *RBM20* downregulation. In resDCM-CMs, the observed Ca^{2+} leakage in DCM-CMs was improved significantly and was comparable to control-CMs. Similar to resLVNC-CMs, the splicing of *RYR2* improved in resDCM-CMs. This is the first report, that RBM20 mutations can cause different pathological phenotypes exemplified by their distinctive

Ca^{2+} handling impairments, which can be attributed to their distinctive RBM20-mutations using isogenic rescue-lines. To fully define these RBM20-mutations as a monogenetic disease cause, control-lines should be artificially modified with these missense mutations. This would help to understand, whether the RBM20 mutations are the sole driver for these sarcomeric and Ca^{2+} handling phenotypes or if unidentified SNPs in the LVNC and DCM patients contribute to the disease progression.

5.7 NOVEL RBM20 SPLICE TARGETS

With control- and rescue-lines available in this study, NGS was performed to screen for novel RBM20 splice targets. For this, the DexSeq pipeline was applied, which allows to quantify the expression of every exon within a gene (performed by the cooperation group AG Meder in Heidelberg) (Anders et al., 2012). This identified a total of 10826 genes in DCM-CMs (vs control/resDCM-CMs) and 9854 in LVNC-CMs (vs control/resLVNC-CMs) with differential exon usage, of which 8947 genes are shared between the LVNC and DCM cohort. From the latter, 36 candidate genes were selected. Primers were designed to span the prospective exons and scanned via gel-electrophoresis. PCR products that showed only one band were discarded. From 36 candidate genes, only six genes showed different isoforms and were subjected to Sanger sequencing and subsequently to qPCR analysis. This underscores that a high number of false positives were deducted from the DexSeq analysis and reassessment of the data should include more stringent cut-off and threshold values. In addition, the identified gene cohorts included many pseudogenes and non-coding transcripts. Exclusion of such transcripts would minimize the number of genes and could benefit the selection of further candidate genes. Here, possible new splice targets of RBM20 are proposed as follows: exon 6 of *MYLK*, exon8/9 of *KCNH2*, exon 13 of *CACNA1C*, exon 14 of *ADCY6*, exon 4 of *BMP7* and exon 10 of *NEXN*. *CACNA1C* was already described for RBM20, however earlier reports in RBM20 knockout rats described exon 9 to be affected (Guo et al., 2012). To our knowledge, exon 13 was identified here for the first time and LVNC- and DCM-CMs show more transcripts harboring exon 13, which suggests that human wt RBM20 is involved in exclusion of this exon. This disparity could be due to inter-species differences as these *CACNA1C*-exon 9 was identified in knockout rats, whereas here human iPSC-CMs were used. Notably, LVNC-CMs showed the most prominent effect that was rescued in resLVNC-iPSC-CMs (Fig. 24B).

KCNH2 is a voltage gated K^+ channel, also known as hERG and has a substantial influence for the cardiac electrophysiology (Oshiro et al., 2010). Here, exon 8/9 was identified to be RBM20 dependent. Although *KCNH2*-exon 8/9 levels are similar in LVNC-, DCM- and control-

CMs, comparison to the rescue-lines showed a clear significance of elevated *KCNH2*-exon 8/9 levels. This insinuates that human wt RBM20 fosters the inclusion of exon 8/9 in *KCNH2*. This is the first report that shows RBM20-dependent splicing of a K⁺-channel. *KCNH2* codes for the Kv11.1 or hERG channel, which shapes phase 3 of the cardiac AP by releasing K⁺ into the extracellular space (Grant, 2009). It is known that patients with exon 7/8 deletion or missense mutations in the *KCNH2* gene show long-QT syndrome (Tester & Ackerman, 2008; Kapa et al., 2009). As defects in KCNH2 are implicated in prolongation of cardiac AP, which is an indicator for arrhythmia, this target gene opens a new avenue for the involvement of RBM20 in ion homeostasis. Arrhythmogenic tendencies in RBM20 null mice has linked RYR2-dependent Ca^{2+} leakage and increased Ca^{2+} currents to this phenotype (van den Hoogenhof et al., 2018). However, AP prolongation by aberrant KCNH2 function could be an additional mechanism by which arrythmia events in RBM20-mutation based cardiomyopathies occur. To analyze the impact of RBM20-dependent missplicing on *CACNA1C* and *KCNH2*, Ca^{2+} and K⁺ currents have to be analyzed in LVNC- and DCM-CMs and compared to resLVNC- and resDCM-CMs.

ADCY6 is a membrane-bound enzyme that produces intracellular cAMP from ATP and is activated by G-protein coupled receptor signaling (Dessauer et al., 2017). Here, ADCY6-exon 14 levels are elevated in DCM-CMs compared to control- and resDCM-CMs. Notably, the ADCY family comprises nine isoenzymes of which ADCY6 and ADCY5 are considered the main enzymes in adult myocytes (Dessauer et al., 2017). Studies in ADCY6 knockdown in HEK293-lines and mice experiments demonstrated no reduction of basal cAMP levels, but reduced cAMP after β-adrenergic stimulation. The mouse model showed reduction of contractile force and SERCA2a performance after β-adrenergic stimulation (Tang et al., 2008; Soto-Velasquez et al., 2018). The influence of ADCY6 and RBM20-dependent splicing remains elusive and further studies of cAMP-levels are needed to assess their influence. Preliminary data from our group of adenovirus-based cAMP measurements in LVNC- and DCM-CMs resulted in elevated basal cAMP in LVNC but no difference after β-adrenergic stimulation (data not shown). Specific ADCY6 inhibitors are needed to analyze the role of ADCY6 in cAMP of the described cardiomyopathies in more detail.

BMP7 is a growth factor that belongs to the TGF-β family and conveys its effects through SMAD signaling pathways. During cardiogenesis, BMP7 signaling is required for the formation of the cardiac cushion area that will develop in the valves and septa (Kim et al., 2001). In DCM-CMs, *BMP7*-exon 4 levels are decreased compared to control- and resDCM-CMs indicating that human wt RBM20 is involved in exon 4 inclusion. To invest this further, Western blot analysis of TGF-β signaling proteins (like SMADs) should be conducted. This would help to understand if the *BMP7* missplicing has functional consequences for its downstream effector signaling.

MYLK-exon 6 and *NEXN*-exon 10 splicing could not be verified on qPCR level. However, the LVNC- and DCM-CMs were used for comparison and a RBM20 knockout line would be necessary to clarify, whether these genes are targets of RBM20.

Here, four novel potential RBM20 splice targets were identified: CACN1c, KCNH2, ADCY6 and BMP7. Of these, LVNC- and DCM-CMs share missplicing phenotypes for *CACNA1c* and *KCNH2*, whereas *ADCY6* and *BMP7* were exclusively affected in DCM-CMs. Notably, additional RBM20 targets exclusively affected in LVNC-CMs have not been found. This data adds to the repertoire of RBM20 splice targets and underscores that the full scope of RBM20-dependent splicing is not fully understood yet. Intriguingly, this is the first report that identifies a developmental gene (BMP7) as a potential target. Regarding RBM20-based LVNC it would be interesting to investigate, whether RBM20 has a prominent role in cardiogenesis. For this, the DexSeq experiment should be repeated with 30d old iPSC-CMs and screened with a emphasis on genes implicated in cardiogenesis.

5.7 LOCALIZATION STUDIES of RBM20 and CAMK2δ

RBM20 is classified as a SR protein, which is a structural classification for splice factors a conserved RS-domain. Reports on other SR protein splice factors identified the RS-domain as a NLS in addition to be crucial for protein-protein interactions (Yeakley et al., 1999). In line with this observation, transient expression of RBM20 constructs in HELA cells showed that RBM20 accumulates in the cytoplasm when the RS-domain and parts of the RRM are deleted (Filippello et al., 2013). Another study complemented this finding by demonstrating that the RS-domain of RBM20 needs to become phosphorylated to function as an NLS (Murayama et al., 2018). Since the RBM20 mutations in LVNC and DCM concern the p.634 position in the RS-domain, the question was raised whether altered distribution can be observed. Immunostainings of RBM20 showed that the protein is detected predominantly in the nucleus with small amounts in the cytoplasm. However, analysis of RBM20 levels in different cell compartments of the nucleus, cytoplasm and membrane-bound fraction yielded conflicting results. Firstly, LVNC- and DCM-CMs showed elevated levels of RBM20 in the nucleus compared to resLVNC-CMs, which is contrary to the literature reports. Secondly, when control-CMs are used for comparison, only LVNC-CMs show a tendency for decreased RBM20 in the nucleus fraction, whereas DCM-CMs are comparable to DCM-CMs (data not shown). This suggests that the results depended on the control group and no conclusive result can be deducted in this study so far. One problem with the analysis was the difficulty to normalize the RBM20 protein amounts. For the Western blot analysis, the whole protein content was used for normalization. This is feasible to correct loading differences of the sample but fails to determine separation qualities in each fraction. Thereof, other localization approaches should be chosen to conclude RBM20 distribution patterns. In addition, analysis of phosphorylation status of the RBM20 RS-domain in LVNC- and DCM-CMs could contribute to understand the consequence of their respective mutations. Such studies should comprise localization studies with ectopic expression of wt and mutated RBM20 in other *in vitro* systems like Hela cells to eliminate the cardiac background bias.

CAMK2δ is a splice target of RBM20, whereas human RBM20 fosters the inclusion of Exon 14/NLS in iPSC-CMs, which is defined as the CAMK2δ_B isoform as described earlier. With an NLS present in CAMK2δ_B, the localization is defined as nuclear (Beckendorf et al., 2018) and the question was addressed, whether altered distribution patterns can be detected as a result of altered splicing. Due to the reduced levels of exon 14/NLS in LVNC-CMs, the hypothesis was tested, whether reduced amounts of CAMK2δ are present in the nucleus of LVNC-CMs. Immunostainings showed localization of CAMK2δ in the cytoplasm and nucleus. As for the study of RBM20 localization, cell fractionation experiments were used. ResLVNC-CMs were chosen as a control group since CAMK2δ-NLS levels were reduced in LVNC-CMs and suggested an altered distribution pattern. Compared to resLVNC-CMs, no differences in CAMK2δ distribution were detected in LVNC- or DCM-CMs in the cytoplasmic, nucleic and membrane fraction. However, the immunostainings for CAMK2δ suggest altered distribution patterns. In resLVNC- and resDCM-CMs, CAMK2δ is detected in the nucleus and the cytoplasm as a diffuse pattern. In contrast, LVNC-CMs show CAMK2δ distribution around the nucleus and distinct patterns in the cytoplasm. As discussed for the RBM20 localization studies, the cell fractionation experiment proved difficult to normalize. Another aspect is the separation into basic cell fractions that does not differentiate for micro-domains in the nucleus or cytoplasm. CAMK2δ isoforms are reported to localize to different cell compartments like the nucleus (CAMK2δ_B isoform), T-tubuli and nuclear membrane (CAMK2δ_A isoform) or the cytoplasm (CAMK2δ_C isoform) (Beckendorf et al., 2018; Gray & Heller Brown, 2014). Therefore, this cell fractionation approach might not be sensitive enough and alternative approaches for the study of CAMK2δ localization in LVNC- and DCM-CMs should be explored. One possibility could be to double-stain CAMK2δ with proteins specific for distinct cell compartments in CMs and analyze co-localization with high-resolution microscopy.

5.8 DEVELOPMENTAL ASSESSMENT of LVNC

LVNC is a rare genetic cardiomyopathy that is characterized by incomplete myocardial compaction and trabecular recesses due to developmental defects (Arbustini et al., 2016). Here, LVNC-CMs at an earlier time point of d35 were investigated for developmental parameters. First, the expression of RBM20 was assessed at d30 and d90 old iPSC-CMs to conclude whether RBM20 is present at early developmental time point. *RBM20* m-RNA showed rapid upregulation after cardiac induction (Fig.30C). Other studies support this observation with RBM20 present in d12 old mouse ESC-CMs and RBM20 induction was shown at E7.5 during murine cardiogenesis (Beraldi et al., 2014). Furthermore, LVNC-CMs showed detained cell growth and decreased proliferation rates (Fig. 30D). Smaller cells in combination with decreased proliferation could contribute to insufficient compaction during development, although the mechanisms by which RBM20 conveys these effects is unclear. A screen for expression of developmental genes showed upregulation of *MEF2C* and *HEY2* in LVNC-CMs. MEF2C is an important cardiac transcription factor in cardiac development and mutations lead to congenital heart diseases (Qiao et al., 2017). Another important pathway in chamber development and trabeculation formation is the NOTCH pathway, which affects among many other genes the expression of *HEY2* (Miao et al., 2018; D'Amato et al., 2016). The conclusive effects of upregulation of *MEF2B* and *HEY2* in LVNC-CMs and their relation to the RBM20 mutation remain speculative. However, this indicates that developmental networks are affected in LVNC-CMs and comprehensive studies at different time points could help to get molecular insights in this scarcely understood cardiomyopathy. Indeed, mouse models of LVNC have been proven difficult, since the phenotype is not completely reiterated and transgenic mice often die *in utero* or at birth (Chen et al., 2009). Here, the need for iPSC-CMs to mature by prolonged culture time presents an advantage, as part of the maturation process is recapitulated *in vitro* and are therefore and excellent system to study LVNC. Since resLVNC-CMs are already available, the influence of the RBM20-R634L variant in LVNC could be assessed directly. To date, only one LVNC-iPSC model was reported. This model investigated the influence of TBX20-Y317* mutation in a patient-specific iPSC-CMs. They also reported reduced proliferation capacity and linked this phenotype to increased TGF-β signaling (Kodo et al., 2016). This suggests proliferation defects as a common trait in LVNC. Whether the decreased proliferation in the RBM20-based LVNC is also influenced by impaired TGF-β signaling should become the focus of subsequent experiments. Our study shows that iPSC-

CMs are suitable for disease-modelling of LVNC. First data demonstrated developmental aberrations in RBM20-based LVNC. However, to conclusively show if RBM20 can cause LVNC, 3D EHM or organoid models are necessary to investigate structural characteristics like compaction and trabeculation.

5.9 FROM BEDSIDE to BENCH and BACK

In this work, one family with RBM20-dependent LVNC and one family with RBM20-dependent DCM donated somatic material for research. The LVNC index patient (II.3) showed intertrabecular recesses and a severely reduced ejection fraction of 20 %. The age of onset/diagnosis was 39 and she received an ICD implantation as a precaution. The ICD revealed multiple non-sustained tachycardia events (Sedaghat-Hamedani et al., 2017). The DCM family has not been published to date, but their underlying RBM20-R634W mutation is a known disease variant for DCM (Li et al., 2010; Hey et al., 2019). In this study, the DCM 1 (III.6) showed an age of onset/diagnosis at 38 with an ejection fraction of 30 %. His brother/DCM 2 (III.7) presented similar symptoms with an early age of onset/diagnosis at 43 and a reduction in ejection fraction with 40 %. Information about tachycardia or arrhythmia events were not available. The study of their patient-specific iPSC-CMs revealed an irregular sarcomeric structure, which could be a main contributor to their reduced heart performance. Furthermore, impaired Ca^{2+} handling has been identified and suggests that treatment with Ca^{2+} channel blockers could benefit these patients in the future. The increased Ca^{2+} leakage identified in the DCM-CMs is an indicator that these patients are at risk for the development of arrhythmia and clinical follow-ups should monitor the patients for possible arrhythmic events.

The study in RBM20-deficient rat and mouse models is limited by inter-species differences. Furthermore, these models investigate the contribution of a gene by knockout studies. However, patients with RBM20 mutations present with heterozygous point mutations that might not resemble a knockout phenotype. In this project, differential disease phenotypes were identified for LVNC and DCM, which depend on RBM20 splice targets. In turn, the RBM20-dependent splicing can segregate into different outcomes depending on the respective mutation. This suggests that disease and risk stratification of patients could be classified by the distinctive RBM20 mutation. Furthermore, the iPSC-CMs system is an excellent platform to screen for potential therapeutic drugs, especially when disease phenotypes have been defined. Due to the prominent Ca^{2+} kinetics phenotype in LVNC-CMs, the question was addressed if therapeutic drugs like Verapamil can improve this disease phenotype. Verapamil is an inhibitor of the L-type Ca^{2+} current (Cav1.2), which decelerates kinetics (Blinova et al., 2017). After 15 min incubation with Verapamil, the Ca^{2+} transient rise time was 1.3-fold increased in LVNC-CMs. Subsequently, stimulation with Iso evoked a prominent response (Fig. 29). This shows that Ca^{2+} kinetics can be modulated with Verapamil and improved the Ca^{2+} phenotype in LVNC-CMs. Due to time constraints, the effect of Verapamil was not tested in DCM-CMs regarding the effect on Ca^{2+} leakage. Verapamil treatment in RBM20 null mice showed a significant reduction in Ca^{2+} leakage (van den Hoogenhof et al., 2018), thereby suggesting that Verapamil could also benefit the DCM-CMs. However, whether Verapamil is a safe option for this RBM20-based LVNC or DCM needs further investigation. One major aspect to consider is that Verapamil also has negative ionotropic (contractility) effects and is contraindicated in the treatment of heart failure. In contrast, Verapamil is tested in the treatment of hypertrophic cardiomyopathy rather than DCM (Lan et al., 2013; Wu et al., 2019). Since the LVNC and DCM patients already suffer from a reduced ejection fraction, it is a concern whether this drug should be administered. To address this issue, Verapamil should additionally be tested in various concentrations in a 3D EHM model of LVNC- and DCM-CMs to assess the mechanical outcome of the drug treatment. Furthermore, a long-term incubation (2-4 weeks) with the prospective drug should test other safety concerns like cardiotoxicity. In addition, long-term treatment could also elucidate whether the treatment has a beneficial effect on the sarcomeric structure, which was another disease aspect in LVNC and DCM. In future studies, other drug candidates should be tested as a treatment option. This includes drugs that can stabilize the RYR2 and therefore benefit Ca^{2+} cycling and leakage (Dantrolene, Rycals) or inhibitors of CAMK2δ (AIP, KN93). In summary, the patient-specific LVNC- and DCM-CMs offer an exquisite platform for personalized medicine and drug screening. Verapamil was identified as a promising candidate in the treatment of LVNC, although further testing is needed.

5.10 LIMITATIONS

The present study used a patient-specific stem cell system, which holds many advantages over animal models. However, several disadvantages and limitations have to be considered. As discussed above, iPSC-CMs resemble fetal/neonatal CMs. Therefore, the phenotypes described in this work could differ in the context of adult CMs. Furthermore, all experiments were performed in the 2D monolayer format, which can not analyze 3D structure phenotypes. Also, the iPSC-CMs were cultured exclusively as ventricular CMs, which does not recapitulate the physiological environment of the heart as a multi-cellular organ.

The study of patient-specific iPSC models is dependent on the willingness of patients to donate somatic material. In this project, a second donor from the LVNC cohort joined later, whereas her iPSC-CMs were only used in the NGS experiments. Also, no healthy family member of the LVNC cohort was available, although the possibility to generate isogenic controls can compensate for this shortcoming.

5.11 CONCLUSION

Fig. 33 summarizes the findings of this study. RBM20-dependent missplicing is proposed as a molecular driver for shared and differential disease phenotypes. While *TTN* missplicing is continuously observed with sarcomeric irregularity in LVNC- and DCM-CMs, differential Ca^{2+} handling impairments can in part be explained by differential missplicing. In particular, *CAMK2δ* missplicing in combination with accelerated Ca^{2+} kinetics and CAMK2δ-dependent hyper-phosphorylation in LVNC-CMs is proposed a main disease mechanism for RBM20-based LVNC. Ca^{2+} channel blockade by Verapamil could prevent some of the disease characteristics, providing a clinical use of Verapamil in patients with RBM20-dependent LVNC, but might not be a sustainable therapeutic option. The functional consequence of *TRDN* missplicing in LVNC-CMs remains to be determined. For DCM, differential missplicing in *LDB3*, *ADCY6* and *BMP7* was observed together with decreased systolic Ca^{2+} and increased Ca^{2+} leakage as typical heart failure characteristics. Nevertheless, the exact mechanism of these pathological phenotypes remains to be elucidated. Isogenic control-lines of LVNC and DCM directly linked these disease phenotypes to the respective RBM20 mutations and demonstrated a rescue of the cardiomyopathy phenotypes. Therefore, a nucleotide specific CRISPR repair was identified as a personalized strategy to treat RBM20-associated LVNC and DCM. Novel splice targets *CACN1C*, *KCNH2*, *ADCY6* and *BMP7* expands the scope of the involvement of RBM20 in cardiac functions and opens a wider avenue for RBM20-dependent disease mechanisms.

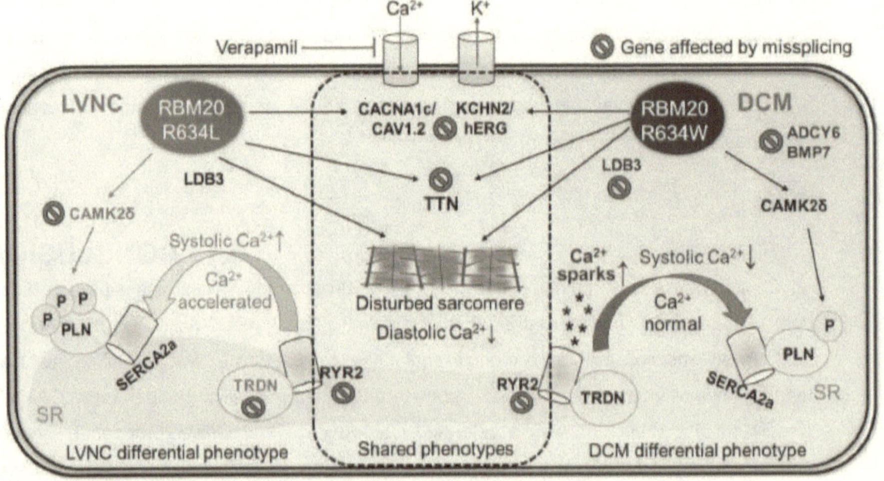

Fig. 33 RBM20-dependent disease mechanisms in LVNC and DCM
LVNC- and DCM-CMs show missplicing in *CACNA1c*, *KCNH2*, *TTN* and *RYR2*. They share disturbed sarcomeric structures and decreased diastolic Ca²⁺ as disease phenotypes. LVNC-CMs show exclusive missplicing in *TRDN* and *CAMK2δ*, which leads to increased systolic Ca²⁺ and accelerated Ca²⁺ cycling by CAMK2δ-dependent hyperphosphorylation of its targets. Treatment with Verapamil ameliorates the Ca²⁺ phenotype in LVNC-CMs. DCM-CMs show exclusive missplicing in *LDB3*, *ADCY6* and *BMP7*. The mechanism for increased Ca²⁺ leakage and decreased systolic Ca²⁺ remain unknown.

5.12 OUTLOOK

This work established a patient-specific iPSC-CM model of RBM20-dependent LVNC and DCM with healthy donors and CRISPR/Cas9 edited rescue-lines serving as control groups. Thereby, shared and differential splicing of RBM20 targets was identified as a molecular disease driver. Cellular pathologies comprised sarcomeric disarray as a shared pathology and differential Ca²⁺ handling impairments in LVNC- and DCM-CMs. CAMK2δ-dependent hyperphosphorylation was identified as a main mechanism in LVNC, which coincided with accelerated Ca²⁺ cycling. In the future it would be interesting to investigate the structural consequence of the observed sarcomeric irregularity phenotype in LVNC- and DCM-CMs in greater detail. For this, myofilament measurements in single cell iPSC-CMs and in a 3D EHM system should be conducted, which could quantify cardiac parameters such as force generation, passive stiffness and Ca^{2+} sensitivity. This would help to understand whether the sarcomeric and Ca^{2+} handling impairments observed, segregate into different diastolic and/or systolic impairments. Furthermore, a 3D model of LVNC could help to identify the role of RBM20 in the disease process. The characteristic of LVNC is a non-compacted structure of the myocardium with intertrabecular recesses. Since this work focused on a 2D monolayer set up for the experiments, such structural aspects could not have been investigated. Therefore, this work cannot conclusively show whether RBM20 is the sole driver for the non-compaction phenotype of LVNC. In addition to iPSC-CMs, the role of cardiac fibroblasts should be assessed. Cardiac fibroblasts have a grave role in extracellular matrix remodeling, a process that is crucial during heart development. We showed that RBM20 is not expressed in endothelial cells, skin fibroblasts and only in small amounts in iPSCs, however whether RBM20 is present in cardiac fibroblasts remains to be determined.

This study defined molecular and cellular disease phenotypes for LVNC- and DCM-CMs, which is an excellent foundation for screening of medical compounds. Due to time constraints, only Verapamil was tested in LVNC-CMs, which showed beneficial effects on Ca^{2+}-kinetics. Other potential candidates should be included in future analysis. Prospects for drug screenings should include treatment of LVNC-CMs with AIP or KN93, an inhibitor of CAMK2δ. For DCM-CMs, treatment with Verapamil should be assessed as well since Verapamil has been shown to reduce Ca^{2+} leakage in an RBM20 mouse model. Further drug candidates for DCM comprise Dantrolene and Rycals since they have been shown to reduce Ca^{2+} leakage by stabilization of the RYR2 receptor. In addition to analyzing the acute effects, follow-up studies should also include a longer incubation phase with the respective compounds. This could help to investigate, whether favorable effects of the drug treatments are sustainable or if cardiotoxic effects occur.

Another aspect of this project was the identification of novel RBM20 splice targets. The identification of *CACNA1C*, *KCNH2*, *ADCY6* and *BMP7* expanded the scope of RBM20 function. As mentioned above, the DexSeq data could be analyzed with more stringent cut-off settings to identify more possible splice targets. In addition, single-cell patch clamp recordings of Ca^{2+} and K^+ currents would aid to decipher the influence of RBM20-dependent mis-splicing of *CACNA1C* and *KCNH2* in disease manifestation.

APPENDIX

6.1 SUPPLEMENTAL FIGURES

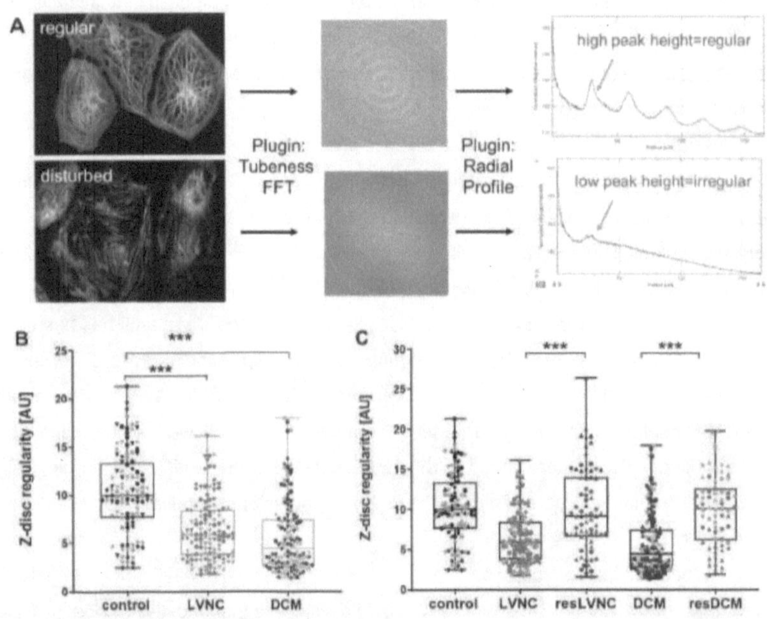

Suppl. Fig. 1 Analysis of sarcomeric regularity
A: Example of sarcomeric assessment using fast Fourier transformation. The height of the first order peak is analyzed for quantification of sarcomeric regularity. Higher values correlate with a more regular structure.
B+C: Quantification of the Z-disc regularity. Dots with the same color and symbol belong to the same cardiac differentiation. The two DCM patients are not separated by color. Data is presented as boxplots with [n = number of differentiations/analyzed cells] for: control (n= 6/112), LVNC (n = 7/128), resLVNC (n = 3/55) , DCM (n = 7/137), resLVNC (n = 4/73) and resDCM (n = 4/69).
B: P-values were calculated by one-way ANOVA multiple comparisons and significances are marked with * p<0.05, ** p<0.01 and *** p<0.001.
C: P-values were calculated by Student's t-test of LVNC vs resLVNC and DCM vs resDCM and significances are marked with * p<0.05, ** p<0.01 and *** p<0.001.

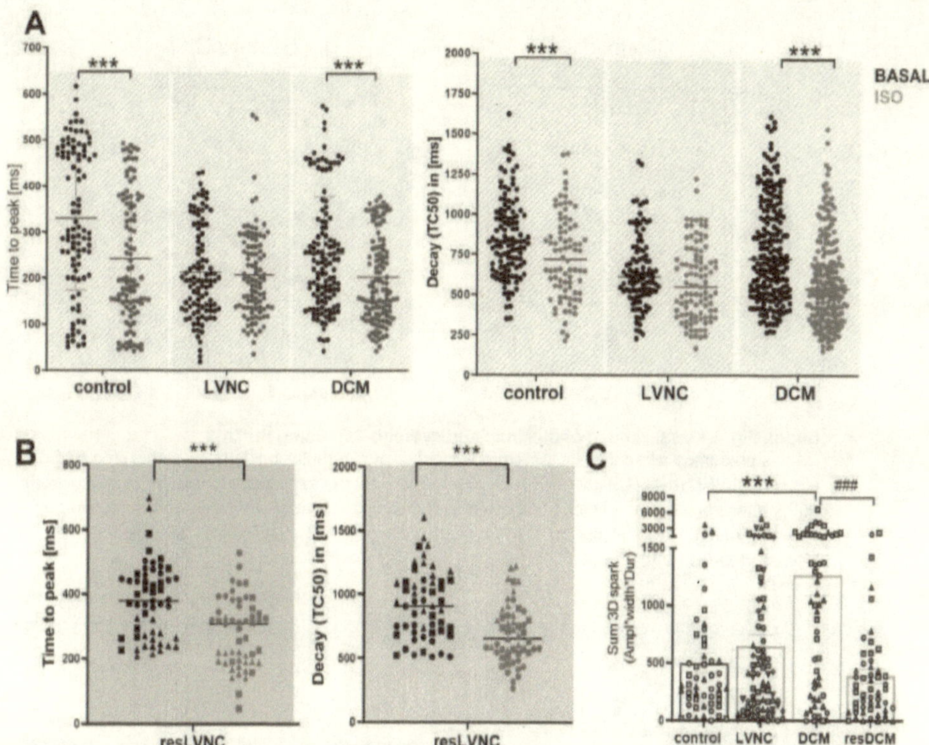

Suppl. Fig. 2 Analysis of Ca²⁺ kinetics and leakage with Fluo-4

A-C: Every dot represents one measured cell.

A: Ca²⁺ transient rise and decay times. Data is presented as scatter blots with [n = number of differentiations/analyzed cells] control (n= 6/94), LVNC (n = 6/86) and DCM (n = 7/126) and Iso stimulated: control (6/101), LVNC (6/99) and DCM (7/119). P-values were calculated by two-way ANOVA (comparing Basal vs Iso) with Sidaks multiple comparisons mixed model and significances are marked with * p<0.05, ** p<0.01 and *** p<0.001.

B: Ca²⁺ transient rise and decay times. Data is presented as scatter blots with [n = number of differentiations/analyzed cells] for basal (black) resLVNC (n = 3/55) Iso stimulated(green) resLVNC (3/53). P-values were calculated by Student´s t-test and significances are marked with * p<0.05, ** p<0.01 and *** p<0.001.

C: Ca²⁺ leakage. Data is presented as bar graphs +/- SEM with [n = number of differentiations/analyzed cells] control (n= 3/52), LVNC (n = 5/82), DCM (n = 3/56) and resDCM (3/53). P-values were calculated by one-way ANOVA against control with Dunnett correction and significances are marked with * p<0.05, ** p<0.01 and *** p<0.001. Additionally, p-values were calculated by Student´s t-test of DCM vs resDCM and significances are marked with # p<0.05, ## p<0.01 and ### p<0.001.

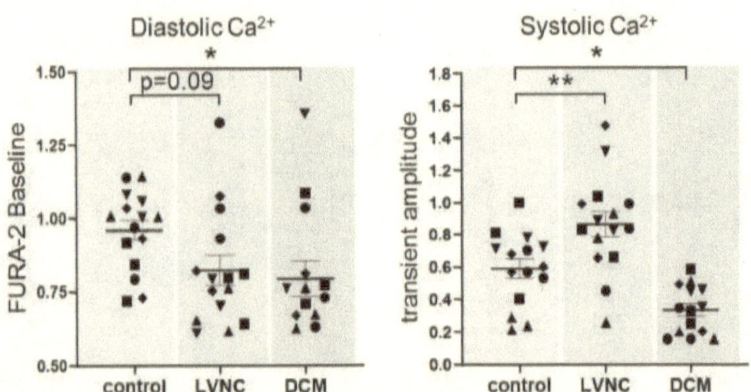

Suppl. Fig. 3 Measurement of diastolic and systolic Ca²⁺ using FURA-2
Data is presented as scatter blots with [n = number of differentiations/analyzed cells] for control
(n= 5/15), LVNC (n = 5/15) and DCM (n = 5/14). Every dot represents one measured cell, whereas
dots with the same shape belong to the same cardiac differentiation. P-values were calculated by
one-way ANOVA against control with Dunnett correction and significances are marked with *
p<0.05, ** p<0.01 and *** p<0.001.

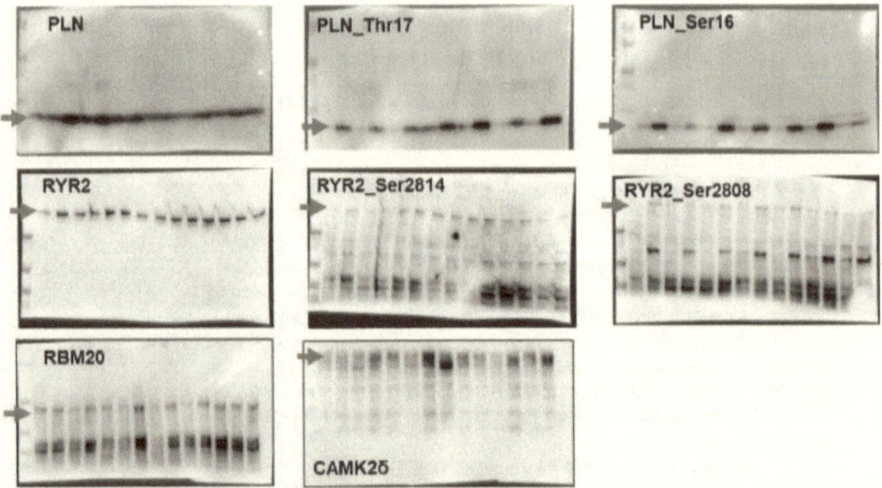

Suppl. Fig. 4 Western blot membranes
The red arrows mark the protein band used for quantification.

6.2 SUPPLEMTNAL TABLES

Suppl. Tab. 1 Primer List. An.: Annealing temperature, Exp.: expression; Ex: Exon, Seq.: sequencing

Gene		Sequence 5´- 3´	An.	Method	bp
Pluripotency Markers					
OCT4	Fwd	GACAACAATGAAAATCTTCAGGAGA	59 °C	RT-PCR	473
	Rev	TTCTGGCGCTTACAGAACCA			
SOX2	Fwd	ATG CAC CGC TAC GAC GTG A	60 °C	RT-PCR	437
	Rev	CTT TTG CAC CCC TCC CAT TT			
NANOG	Fwd	AGTCCCAAAGGCAAACAACCCACTTC	62 °C	RT-PCR	164
	Rev	ATCTGCTGGAGGCTGAGGTATTTCTGTCTC			
LIN28	Fwd	AGTAAGCTGCACATGGAAGG	52 °C	RT-PCR	410
	Rev	ATTGTGGCTCAATTCTGTGC			
GDF3	Fwd	TTCGCTTTCTCCCAGACCAAGGTTTC	54 °C	RT-PCR	311
	Rev	TACATCCAGCAGGTTGAAGTGAACAGCACC			
FOXD3	Fwd	GTGAAGCCGCCTTACTCGTAC	61 °C	RT-PCR	258
	Rev	CCGAAGCTCTGCATCATGAG			
Cardiac markers					
cTNT	Fwd	GACAGAGCGGAAAAGTGGGA	62	RT-PCR	305
	Rev	TGAAGGAGGCCAGGCTCTAT			
RBM20 exp.	Fwd	CCTCCACTTGCCGCATATCTGT	60	qPCR	107
	Rev	AGACCAGGCATTTCTGAGCGTG			
RBM20 seq.	Fwd	GAGTGTACACAGTTACATGCAC	56	RT-PCR	180
	Rev	GTGGGACCTCGGGGAGA			
α-MHC	Fwd	GTCATTGCTGAAACCGAGAATG	58	RT-PCR	413
	Rev	GCAAAGTACTGGATGACACGCT			
Germlayer markers for mass culture					
AFP	Fwd	ACTCCAGTAAACCCTGGTGTTG	60	RT-PCR	255
	Rev	GAAATCTGCAATGACAGCCTCA			
ALB	Fwd	CCTTTGGCACAATGAAGTGGGTAACC	60	RT-PCR	355
	Rev	CAGCAGTCAGCCATTTCACCATAGG			
MAP2	Fwd	CCACCTAGAATTAAGGATCA	55	RT-PCR	482
	Rev	GGCTTACTTTGCTTCTCTGA			
Housekeeping genes					
GAPDH	Fwd	AGAGGCAGGGATGATGTTCT	60	qPCR	258
	Rev	TCTGCTGATGCCCCCATGTT			
18s	Fwd	ACCCGTTGAACCCCATTCGTGA	60	qPCR	150
	Rev	GCCTCACTAAACCATCCAATCGG			
RBM20 splice targets					
RYR2 exp.	Fwd	CTTGAGGTTGGCTTTCTGCCAG	60	qPCR	153
	Rev	TGTGCCAGCAAAGAGAGGAGCA			
RYR2-24bp insertion	Fwd	GTCACAGGATCCCAACGCAG	60	qPCR	124
	Rev	CTTTGCTGGCACTGATTGTCTG			

CACNA1C exp.	Fwd	GCAGGAGTACAAGAACTGTGAGC	60	qPCR	143
	Rev	CGAAGTAGGTGGAGTTGACCAC			
CACNA1C ex9	Fwd	CGAAGGCATGGATGAGGAGA	60	qPCR	99
	Rev	CCCTCGATGTCACCTCCAGC			
TRDN exp.	Fwd	GGAGGACAAAGAGAAAGCAGCTG	60	qPCR	120
	Rev	AGGTGGAATGGCTGGGCTTTGT			
TRDN ex9	Fwd	GTCCATGGGGATTTAAAACCAGG	60	qPCR	100
	Rev	CTTCAAGGGCAGGTGATGC			
CAMK2δ exp.	Fwd	ACACGGTGACTCCTGAAGCCAA	60	qPCR	145
	Rev	GTCTCCTGTCTGTGCATCATGG			
CAMK2δ ex14/NLS	Fwd	CCATCTTGACAACTATGCTGGCT	60	qPCR	149
	Rev	GAACACTCGAACTGGACTTCCT			
CAMK2δ ex15/16	Fwd	AAGATGTGAAAGCACGAAAGC	60	qPCR	88
	Rev	GTGTAGGCTTCAAAGTCCCCAT			
CAMK2γ exp.	Fwd	ATCCACGGTGGCATCCATGA	60	qPCR	107
	Rev	AGACAAGCATGGTCGTGAGG			
CAMK2γ ex14/NLS	Fwd	AACAAGAAGTCGGATGGCGG	60	qPCR	129
	Rev	CATGGCCGTCTGCAAGG			
LDB3 exp.	Fwd	ACCTCGTGGTGGCCATTG	60	qPCR	133
	Rev	GTGGAGATGGGAATGGGACG			
LDB3 ex5	Fwd	TCAAAGCGTCCCATTCCCATC	60	qPCR	197
	Rev	CGG GAG AAG GCA GGG CTA AA			
SORBS1 exp.	Fwd	TATCAGCCTGGCAAGTCTTCCG	60	qPCR	390
	Rev	CCCGTCTGATTCCCTCTTCACT			
SORBS1 ex5	Fwd	GCA CGC TCT ATT TCT GCT GT	60	qPCR	205
	Rev	CTG GTT TGC TTT CGT GTT GCC			
TTN N2B ex50-219	Fwd	CCAATGAGTATGGCAGTGTCA	60	qPCR	92
	Rev	TACGTTCCGGAAGTAATTTGC			
TTN N2BA Ex50-51	Fwd	GCCACAGTCACTGTGACAGAGG	60	qPCR	84
	Rev	GGCTGCCTTACCCACAAAAG			
TRDN ex7-10	Fwd	CAGAAACAAAGACACTGGCG	55	RT-PCR	
	Rev	CTTCAAGGGCAGGTGATGC			
LDB3 ex4-7	Fwd	CCCATTCCCATCTCCACGAC	56	RT-PCR	766
	Rev	GAGACTGCAGGTTGGAGGAA			
SORBS1 ex4-6	Fwd	GGTGGTTCCAAAGCTGTGATG	56	RT-PCR	281
	Rev	GGTTTGCTTTCGTGTTGCCG			
CAMK2δ ex11-15	Fwd	ACTAAAGGGTGCCATCTTGACA	55	RT-PCR	209
	Rev	AGGCTTCAAAGTCCCCATTGT			
Camk2γ ex10-18	Fwd	ATCCACGGTGGCATCCATGA	56	RT-PCR	514
	Rev	GGCCTCAAAGTCCCCATTGT			
Novel RBM20 splice targets					
RBM24 ex1-4	Fwd	GCGAAAGGGTGCGGGAG	57	RT-PCR	862
	Rev	TGGAAAATTGGGAGGGAAAGAGA			
RBM20 ex2-8	Fwd	CGGGGGTCGGCTTAACAAC	58	RT-PCR	766
	Rev	CCTCTCCCTCTGGGAATGGAT			

RBM20	Fwd	TCATCCAGGACATCCATTCCC	58	RT-PCR	860
ex8-11	Rev	TGCCTCACTTTCACTCTCCCA			
REPS2	Fwd	GTGCAAAGCGGGTTGGTTATT	57	RT-PCR	363
ex2-4	Rev	TGCTCCGCAGTTTGCTGTAG			
MYLK	Fwd	TTCAGGACGCCTTTCCTGTC	56	RT-PCR	808
ex4-9	Rev	AGTTCAGCTGACATCGAGGC			
MYLK	Fwd	AGGAAGGCAGCATTGAGGTT	56	RT-PCR	554
ex12-15	Rev	AGTGACCTGGCTTCCATCCA			
MYLK3	Fwd	GTAGTGAGCGTCAAGGAGACC	58		875
ex5-12	Rev	TTCTGCAGCAGTAGTTGGGATT		RT-PCR	
ADAM10	Fwd	ACGTCTAGATTTCCATGCCCA	57		945
ex2-9	Rev	TTGGGAGGTACATGAGACCCA			
ADAM17	Fwd	CGGAACACTTCATGGGATAATGC	58	RT-PCR	496
ex7-11	Rev	CCTCATTCGGGGCACATTCT			
CALM1	Fwd	CTGATCAGCTGACCGAAGAAC	59	RT-PCR	415
ex2-5	Rev	TCATAGTTGACTTGTCCGTCTCC			
CALM3	Fwd	CTGACCAGCTGACTGAGGAG	59	RT-PCR	393
ex2-5	Rev	CATCGATGTCAGCCTCCCTG			
CUL3	Fwd	TCCAGCGTAAGAATAACAGTGGT	57	RT-PCR	852
ex2-7	Rev	GAGCTTTACCTTGCTCCCTCA			
MCU	Fwd	AGACCTCCTCCTCCTTGATGA	56	RT-PCR	511
ex4-7	Rev	ACGTGACTTTTTGGCTCCTTT			
NOTCH2	Fwd	ACTGGGAGCTACTGTGAGGA	57	RT-PCR	689
ex21-25	Rev	GGACTGGGGCAGAAGCAG			
BMP4	Fwd	TAACCGAATGCTGATGGTCGT	57	RT-PCR	1105
ex3-4	Rev	ACAGGCTTTGGGGGATACTGG			
SRSF3	Fwd	GATTTGAGCCGCCGCATTTT	57	RT-PCR	509
ex1-5	Rev	TCGACTACGAGACCTAGAGAAG			
SF3B1	Fwd	CCTCCTCCAGCTGGTTATGT	57	RT-PCR	967
ex1-5	Rev	GTTGGCGGATACCCTTCCAT			
SMURF1	Fwd	ACCATTAGCGTGTGGAACCAT	57	RT-PCR	555
ex4-8	Rev	CGTGCTAACTCCAGTCTGTGT			
ADCY5	Fwd	CTACGTGCTCATCGTGGAGG	58	RT-PCR	862
ex15-20	Rev	GTTCACGGTATTGCCCCAGA			
ADCY6	Fwd	ACCTGCAGCATGAGAATCGG	59	RT-PCR	650
ex2-7	Rev	GTGCTGCTCCTTGAGGTACG			
ADCY6	Fwd	CTGCTGCTGCTAATCACCGT	59	RT-PCR	729
ex12-18	Rev	CAGCAGCCTCCGGTTGTATG			
HDAC9	Fwd	GCATGAGAACTTGACACGGC	57	RT-PCR	541
ex2-6	Rev	TCCATCCTTCCGCCTGAGTA			
HDAC9	Fwd	CCTGAGCATGCTGGACGAAT	57	RT-PCR	330
ex13-16	Rev	TTTGGAAGCCAGCTCGATGA			
ABCC9	Fwd	TCTGGGGCAGATCAACAACTTA	58	RT-PCR	527
ex8-12	Rev	GTGGTGTGACCAGGATATGGA			
ABCC9	Fwd	CACCTGGACAGCTATGAGCA	57	RT-PCR	221
ex14-16	Rev	GTCTGCATCTCACCGAGGAT			
ABCC9	Fwd	AGAATGGATGGGTCTCACAGC	58	RT-PCR	584

ex25-30	Rev	GACCCAAGCCTACCAATCCA			
LMNA ex1-6	Fwd	CGAGGTGTCCGGCATCAA	57	RT-PCR	909
	Rev	GCGTGGATCTCCATGTCCAG			
KCNQ1 ex3-7	Fwd	CTTTGCCCGGAAGCCCATTT	57	RT-PCR	381
	Rev	ACCTTGTCCCCATAGCCGAT			
KCNQ3 ex3-10	Fwd	TCTTTGGAGCCGAGTTTGCT	57	RT-PCR	795
	Rev	GAACCCGATCCAAGAGACCC			
KCNQ1 ex7-9	Fwd	CCATCGGCTATGGGGACAAG	59	RT-PCR	292
	Rev	CTGGGTGACAGCAGAGTGTG			
KCNQ1 ex10-13	Fwd	CTCACAGTCCCCCATATCACG	59	RT-PCR	323
	Rev	TACTGCTCAATGACGTCCCG			
KCNH2 ex7-10	Fwd	AGTATGTGACGGCGCTCTAC	57	RT-PCR	732
	Rev	ACCAGAAGTGGTCGGAGAAC			
KCNH2 ex 12-15	Fwd	AGCCCCTGATGGAGGACTG	57	RT-PCR	481
	Rev	CCTTCTTGGGGAAGCTCTGG			
KCNE1 ex3-4	Fwd	TCTCTGGCCAGTTTCACACAAT	58	RT-PCR	583
	Rev	AAATCAGGTTGCCAGGCAGGAT			
KCND3 ex2-4	Fwd	CATGGTGCCTAAGACGATTGC	58	RT-PCR	317
	Rev	GATGATGCTGGCTCTCGATG			
KCNA4 ex1-4	Fwd	CAGCCATTTCCGAAAACCACC	58	RT-PCR	1987
	Rev	TTCACACATCAGTCTCCACAGC			
SCN5A ex13-16	Fwd	CAGTCAGCGTCCTCACCAG	57	RT-PCR	547
	Rev	GCTAGCACCAGTGTCAGGTT			
ATP2A2 ex8-11	Fwd	TTGGGCACTTCAATGACCCG	57	RT-PCR	520
	Rev	ACTAGGCAAGTGAGAGCAGTC			
RYR2 ex3-9	Fwd	AGACCTCTCCATCTGCACCT	58	RT-PCR	491
	Rev	CTGCCTCACTTCCTGAGCTG			
RYR2 ex16-19	Fwd	GCATAGACCGTTTGCACGTC	57	RT-PCR	450
	Rev	GGTTCACAAGACGTGTCTGC			
RYR2 ex21-26	Fwd	ATCGATGGCCTCTTCTTTCCA	57	RT-PCR	660
	Rev	TGCATTTTCTGCCAACTTGTCC			
RYR2 ex26-29	Fwd	CAATGGTGGACAAGTTGGCA	57	RT-PCR	599
	Rev	CTTGAAAGCCAGTTCTGAGCC			
RYR2 ex29-33	Fwd	GGTGGCATCAGGGCAATGAA	57	RT-PCR	908
	Rev	AATCCAGCCCACCCAGACAT			
RYR2 ex45-47	Fwd	GCTGGTTGTGGACTGCAAAG	57	RT-PCR	398
	Rev	CAGCGTCCCAAGAGGTCAAT			
RYR2 ex50-53	Fwd	GAGTGCTACAGACATGGCCT	57	RT-PCR	605
	Rev	AGGTATCAACAGGTTGTGGGT			
RYR2 ex57-63	Fwd	TTTCTGTGGACGCTGCCCAT	58	RT-PCR	625
	Rev	AGTGAAATCCTATGCCTGACAAGA			
OBSCN ex28-33	Fwd	GGGCAGGCAAGACGATGG	58	RT-PCR	1015
	Rev	CAGGACAGGACCACTGACTG			
OBSCN ex56-58	Fwd	GGAGGCACCGCACACTTATG	59	RT-PCR	727
	Rev	CCAAATGGAAGGAGACAGTGC			
OBSCN	Fwd	GTGACCTGCACCTACTGTGG	58	RT-PCR	1369

ex65-67	Rev	ACCACGGCTGTGATGATGAG			
OBSCN ex70-75	Fwd	CCTGCTCCCAGATTATACTGGTT	57	RT-PCR	665
	Rev	CCACCAAGGTCAGCAAAACT			
MYBPC3 ex2-12	Fwd	CCTGCCGACCAGGGATCTTA	59	RT-PCR	764
	Rev	TGCCGTAGGATCTCCCACAC			
MYBPC3 ex21-26	Fwd	GACACCATTGTGGTTGTAGCTG	57	RT-PCR	760
	Rev	GGCAGTACTCCACGCTGTA			
NEXN ex8-12	Fwd	GGTACCGCCCTGGTAAACTC		RT-PCR	925
	Rev	TCTGGTTTGGGTTCTCCTGT			
CACNA1c ex10-15	Fwd	GTCCGTCAACACCGAAAACG	57	RT-PCR	746
	Rev	CGCCATAAGCCATGATCCCA			
BMP7 ex1-7	Fwd	GGGCTTCTCCTACCCCTACAA	59	RT-PCR	919
	Rev	GAGGACGGAGATGGCATTGAG			
CACNA1c ex13	Fwd	CAACACGCTCACCATTGCCT	60	qPCR	179
	Rev	CGACGAAGCAGTCAAAGCGG			
MYLK ex6	Fwd	AAAGAAGGAGCCACCGCCAA	60	qPCR	139
	Rev	GGCTGAAAGTCCCCCGGAT			
MYLK exp.	Fwd	GAGGTGCTTCAGAATGAGGACG	60	qPCR	127
	Rev	GCATCAGTGACACCTGGCAACT			
KCNH2 ex7/8	Fwd	CATCTGCGTCATGCTCATTGGC	60	qPCR	105
	Rev	TCTGGTGGAAGCGGATGAACTC			
KCNH2 exp.	Fwd	CATCGCCGTCCACTACTTCA	60	qPCR	116
	Rev	CAGTCTTCAGCAGCCCGAT			
ADCY6 ex14	Fwd	CACATAGCACCGCAGTTGGCAT	60	qPCR	141
	Rev	AGGCAGTGATGTCAGCAGGTGT			
ADCY6 exp.	Fwd	CGGGAAGTTGGCCATGATCTTT	60	qPCR	131
	Rev	TCTCATTGGAAGAAGCCAAGCC			
BMP7 ex4	Fwd	CAGAACCGCTCCAAGACGCC	60	qPCR	115
	Rev	AGCTGACATACAGCTCGTGCTTC			
BMP7 exp.	Fwd	CGGGAACGCTTCGACAATGAG	60	qPCR	141
	Rev	GATGTCAAACACCAGCCAGCC			
NEXN exp.	Fwd	GCGTTTGCTGAAGCAAGGAGA	60	qPCR	166
	Rev	TCCGTTCCTCCTCTGTTCGT			
NEXN ex10	Fwd	AAGAGCGAGCAAGAAGGAGAGC	60	qPCR	
	Rev	GTGAGTAAATGGAGCCTCGCTTT			

Suppl. Tab. 2 Mean Cq-values.

Gene	Cq-value	Gene	Cq-value
18s exp.	16	*LDB3* ex5	26
ADCY6 ex14	27	*LDB3* exp.	21
ADCY6 exp.	26	*MEF2B* exp.	28
BMP7 ex4	26	*MYLK* ex6	34
BMP7 exp.	28	*MYLK* exp.	33
CACNA1C ex13	28	*NEXN* ex10	24
CACNA1C ex9	27	*NEXN* exp.	24
CACNA1C exp.	26	*RBM20* exp.	24
CAMK2δ ex14/NLS	25	*RYR2* exp.	21
CAMK2δ ex15/16	24	*RYR2*-24bp insertion	24
CAMK2γ ex14/NLS	28	*SORBS1* ex5	23
CAMK2δ exp.	24	*SORBS1* exp.	22
CAMK2γ exp.	28	*TRDN* ex9	31
GAPDH exp.	21	*TRDN* exp.	27
HEY2 exp.	26	*TTN* N2B ex50-219	25
KCNH2 ex8/9	27	*TTN* N2BA Ex50-51	19
KCNH2 exp.	27		

Suppl. Tab. 3 Cell lines and differentiation ID-numbers used in experiments. R = RBM20

Experiment	control	LVNC	resLVNC	DCM	resDCM
QPCR for RBM20 splice-targets (Fig. 13, Fig. 20 and Fig. 23)	Ctl1_338 Ctl1_349 Ctl1_556 Ctl2_310 Ctl3_119 Ctl3_168 Ctl4_48 3R8_104 4R2_122	6R1_129 6R1_155 6R1_175 6R1_132 6R1_183 6R1_217 6R2_232 6R1_254 6R2_272	6cr11_10 6cr11_13 6cr11_17 6cr20_38 6cr20_47 6cr29_16 6cr29_63	9R_25 9R2_46 9R2_50 9R4_118 9R4_175 9R4_66 10R3_11 10R3_19 10R3_43 10R5_22 10R5_43	9cr35_14 9cr35_20 9cr49_17 9cr49_20 9cr49_96
Sarcomere regularity analysis (Fig. 14 and Fig. 21)	Ctl1_349 Ctl1_356 Ctl1_444 Ctl4_23 4R2_202 4R6_162	6R1_129 6R1_132 6R1_158 6R1_168 6R2_87 6R2_232 6R2_241	6cr11_6 6cr11_10 6cr21_30 6cr29_16	9R2_46 9R2_50 9R4_59 9R4_69 10R3_19 10R5_19 10R5_22	9cr35_14 9cr35_20 9cr49_14 9cr49_20
Ca²⁺ kinetics with Fluo-4 (Fig. 16 and Fig. 21)	Ctl1_349 Ctl3_71 Ctl4_23 Ctl4_28 WTD2_879 4R2_202	6R1_158 6R1_168 6R1_175 6R2_153 6R2_229 6R2_232	6cr11_6 6cr11_17 6cr21_20	9R2_84 9R4_56 9R4_66 9R4_118 10R3_43 10R3_45 10R5_45	-

Ca^{2+} kinetics after Verapamil (Fig. 28)	-	6R1_341 6R1_353 6R1_367 6R1_370	-	-	-
Ca^{2+} sparks with Fluo-4 (Fig. 16 and Fig. 21)	Ctl1_543 WTD2_1108 4R2_202	6R1_158 6R1_175 6R2_153 6R2_229 6R2_241	-	9R4_66 10R3_43 10R5_45	9cr35_14 9cr35_20 9cr49_20
Ca^{2+} load with FURA-2 (Fig. 17)	Ctl1_291 Ctl1_543 WTD2_1151 WTD2_1801 4R2_202	6R1_251 6R1_254 6R2_241 6R2_272 6R2_272	-	9R2_145 9R4_162 10R3_138 10R5_155 10R5_164	-
Multi-electrode assay (MEA) (Fig. 18 and Fig. 21)	Ctl1_499 Ctl1_541 Ctl1_543 Ctl3_71 WTD2_1108 4R2_280	6R1_125 6R1_132 6R1_158 6R1_168 6R1_251 6R1_254 6R2_272	6cr11_17 6cr11_83 6cr20_35 6cr21_20	9R2_84 9R4_59 10R3_138 10R5_19 10R5_155	9cr35_17 9cr49_20 9cr49_96 9cr49_115
Next-generation sequencing (NGS)	Ctl1_356 Ctl2_332 Ctl3_71 Ctl4_16 4R2_202 4R6_162	6R1_125 6R1_129 6R2_66 8R1_62 8R3_13 8R3_99	6cr11_10 6cr11_13 6cr11_17 6cr21_20 6cr29_16	9R2_46 9R2_50 9R4_59 10R3_19 10R5_19 10R5_22	9cr35_14 9cr35_20 9cr35_42 9cr49_17 9cr49_20
Single-cell sequencing (SCS)	-	6R1_357	6cr11_100	9R4_107	9cr49_107
Western Blot for RYR2 and PLN (Fig. 24 and Fig. 25)	-	6R1_125 6R1_129 6R1_175 6R2_153	6cr11_13 6cr11_17 6cr29_16	9R4_59 9R4_118 10R3_11 10R3_43	9cr35_14 9cr35_20 9cr49_20
EdU proliferation assay (Fig. 29)	Ctl1_543 4R2_302 4R2_321 4R6_265	6R1_248 6R1_254 6R1_334 6R2_266 6R2_272	6cr11_124 6cr20_80 6cr29_84 6cr29_169	9R2_158 9R4_175 9R4_175 9R4_263 9R4_265	9cr35_9 9cr35_91 9cr35_96 9cr49_96 9cr49_167
Cell surface (Fig. 29)	Ctl1_356	6R1_129	-	10R5_22	-
QPCR of developmental genes (Fig. 29)	Ctl1_342 Ctl1_349 Ctl1_543 Ctl3_168 Ctl3_170 Ctl4_23 4R2_202 4R6_126	6R1_129 6R1_175 6R1_254 6R2_241 6R2_272	-	9R2_50 9R2_158 9R4_59 9R4_175 10R3_19 10R5_22 10R5_43 10R5_53	-
Western blot of cell fractions (Fig. 26 and Fig. 27)	-	6R1_158 6R1_175 6R1_248 6R1_254 6R2_232	6cr11_10 6cr11_13 6cr11_17 6cr21_20 6cr29_16	9R4_59 9R4_175 10R3_43 10R5_43	-

Suppl. Tab. 4 List of materials.

Material	Company	Order number
96 well plate clear	Brand	781602
96 well plate qPCR	Biorad	HSP9601
Aluminum foil	Roth	2596.1
Cell scraper	Sarstedt	83.1830
Cover slips round 20 mm	Menzel	CB00200RA1
Cover slips round 25 mm	VWR	ECN 631-1584
Cryotube	Thermo Fisher Scientific	377224
Culture dish 12 well plate	CytoOne	CC7682-7512
Culture dish 6 cm	Sarstedt	82.1194.500
Culture dish 6 well plate	CytoOne	CC7682-3359
FLOW tubes	BD Falcon	352058
Fluoro dish 35 mm	WPI	FD35-100
Glass bottles 50 ml, 250 ml, 500 ml, 1000ml	Schott	Order form
MEA plate 6-well	Multi Channel System	Custom made
Nitrocellulose membrane	Sigma-Aldrich	GE10600002
Pacer electrode and chamber	Custom made	Custom made
Pasteur pipettes	Brand	747715
PCR stripes 8x	Sarstedt	72.911.002
Pipette filter tips 0.1 – 1000 µl	Starlab	#S1120-3810,#S1122-1830, #S1120-1840
Pipette tips 0.1 – 1000 µl	Sarstedt	#S1111-3700, #S1111-1706, #S1112-1720
Pipettes 5 ml, 10 ml, 25 ml	Sarstedt	#86.1253.001, #86.1254.001, #86.1685.001
PVDF membrane	Sigma-Aldrich	GE10600023
Scalpel No. 22	Feather	02.001.30.022
Sterile filter Steriflip 50 ml	Merck	SCGP00525
Sterile filter Steritop 500ml	Merck	SCGPT05RE
Tube 0.5 ml	Eppendorf	0030 121.023
Tube 1.5 ml	Eppendorf	0030 120.086
Tube 15 ml	Starlab	E1415-0700
Tube 2.0 ml	Eppendorf	0030 120.094
Tube 5.0 ml	Labsolute	7696741
Tube 50 ml	Starlab	E1450-0700
Whatmann blotting paper	Sigma-Aldrich	WHA10426890
Wipes precision and task	Kimtech	#05511, #7558

Suppl. Tab. 5 List of equipment.

Equipment	Company
Agarose gel chamber	Biorad
Balance EWJ	Kern
Centrifuge table top	Roth
Centrifuges: 5810R, 5415D, 5415R	Eppendorf
Counting chamber Neubauer improved	Marienfeld Superior
Electrophoresis chambers: Mini-PROTEAN Tetra Vertical Electrophoresis Cell	Biorad
FACS Canto II	BD Bioscience
Freezer -20 °C and -80 °C	#Liebherr, #Sanyo
Heat sterilization	Memmert
Heatblock Thermomixer comfort 1.5 ml	Eppendorf
Ice boxes	Neolab
Imager ChemiDoc XRS	Biorad
Incubator HERACELL	Thermo Fisher Scientific
LSM 710 confocal microscope system	Carl Zeiss
Microscopes: Axio Oberserver A1, Axio Oberserver Z1, Primo Vert, Axiovert 25	Carl Zeiss
Microwave	AEG
Milli-Q water preparation system	Merck
Mr. Frosty freezing Box	Thermo Fisher Scientific
Multi-electrode MEA2100	Multichannel systems
Myopacer ES	IonOptix
NanoDrop	Implen
Olympus Motic AE31 microscope	Motic
PCR cycler Thermocycle 48	Sensoquest
pH meter	InoLab
Pipette Accu-jet Pro	Brand
Pipettes; one channel (0.1 – 1000 µl)	Eppendorf
Power supply	Biorad
qPCR CFX connect Real-time system	Biorad
Refrigerator	Liebherr
Rotary shaker	Heto
Shaker Polymax 1040	Heidolph
Sterile work bench HERASAFE 2030i	Thermo Fisher Scientific
Spectrometer	Bio-Tek
Transfer System Trans-Blot Turbo	Biorad
Vortex Neomix 7-2020	Neolab
Waterbath	Memmert

7

REFERENCES

REFERENCES

Abdallah, A.M., Carlus, S.J., Al-Mazroea, A.H., Alluqmani, M., Almohammadi, Y., Bhuiyan, Z.A., and Al-Harbi, K.M. (2019). Digenic Inheritance of LAMA4 and MYH7 Mutations in Patient with Infantile Dilated Cardiomyopathy. Medicina (Kaunas, Lithuania) *55*.

Agarkova, I., Schoenauer, R., Ehler, E., Carlsson, L., Carlsson, E., Thornell, L.-E., and Perriard, J.-C. (2004). The molecular composition of the sarcomeric M-band correlates with muscle fiber type. European journal of cell biology *83*, 193–204.

Ahmed, R.E., Anzai, T., Chanthra, N., and Uosaki, H. (2020). A Brief Review of Current Maturation Methods for Human Induced Pluripotent Stem Cells-Derived Cardiomyocytes. Frontiers in cell and developmental biology *8*, 178.

Albakri, A. (2018). Dilated cardiomyopathy: a review of literature on clinical status and meta-analysis of diagnostic and clinical management. Journal of Clinical Investigation and Studies *1*.

Ambrosi, C.M., Sadananda, G., Han, J.L., and Entcheva, E. (2019). Adeno-Associated Virus Mediated Gene Delivery: Implications for Scalable in vitro and in vivo Cardiac Optogenetic Models. Frontiers in physiology *10*, 168.

Anders, S., Reyes, A., and Huber, W. (2012). Detecting differential usage of exons from RNA-seq data. Genome research *22*, 2008–2017.

Arbustini, E., Favalli, V., Narula, N., Serio, A., and Grasso, M. (2016). Left Ventricular Noncompaction: A Distinct Genetic Cardiomyopathy? Journal of the American College of Cardiology *68*, 949–966.

Arimura, T., Inagaki, N., Hayashi, T., Shichi, D., Sato, A., Hinohara, K., Vatta, M., Towbin, J.A., Chikamori, T., Yamashina, A., and Kimura, A. (2009). Impaired binding of ZASP/Cypher with phosphoglucomutase 1 is associated with dilated cardiomyopathy. Cardiovascular research *83*, 80–88.

Barbosa-Morais, N.L., Irimia, M., Pan, Q., Xiong, H.Y., Gueroussov, S., Lee, L.J., Slobodeniuc, V., Kutter, C., Watt, S., Colak, R., Kim, T., Misquitta-Ali, C.M., Wilson, M.D., Kim, P.M., Odom, D.T., Frey, B.J., and Blencowe, B.J. (2012). The evolutionary landscape of alternative splicing in vertebrate species. Science (New York, N.Y.) *338*, 1587–1593.

7 REFERENCES

Barrangou, R., and Doudna, J.A. (2016). Applications of CRISPR technologies in research and beyond. Nature biotechnology *34*, 933–941.

Beckendorf, J., van den Hoogenhof, M.M.G., and Backs, J. (2018). Physiological and unappreciated roles of CaMKII in the heart. Basic Research in Cardiology *113*, 29.

Behere, S.P., and Weindling, S.N. (2015). Inherited arrhythmias: The cardiac channelopathies. Annals of pediatric cardiology *8*, 210–220.

Ben Jehuda, R., Shemer, Y., and Binah, O. (2018). Genome Editing in Induced Pluripotent Stem Cells using CRISPR/Cas9. Stem cell reviews and reports *14*, 323–336.

Beraldi, R., Li, X., Martinez Fernandez, A., Reyes, S., Secreto, F., Terzic, A., Olson, T.M., and Nelson, T.J. (2014). Rbm20-deficient cardiogenesis reveals early disruption of RNA processing and sarcomere remodeling establishing a developmental etiology for dilated cardiomyopathy. Human molecular genetics *23*, 3779–3791.

Bers, D.M. (2008). Calcium cycling and signaling in cardiac myocytes. Annual review of physiology *70*, 23–49.

Bilic, J., and Izpisua Belmonte, J.C. (2012). Concise review: Induced pluripotent stem cells versus embryonic stem cells: close enough or yet too far apart? Stem cells (Dayton, Ohio) *30*, 33–41.

Blinova, K., Stohlman, J., Vicente, J., Chan, D., Johannesen, L., Hortigon-Vinagre, M.P., Zamora, V., Smith, G., Crumb, W.J., Pang, L., Lyn-Cook, B., Ross, J., Brock, M., Chvatal, S., Millard, D., Galeotti, L., Stockbridge, N., and Strauss, D.G. (2017). Comprehensive Translational Assessment of Human-Induced Pluripotent Stem Cell Derived Cardiomyocytes for Evaluating Drug-Induced Arrhythmias. Toxicological sciences an official journal of the Society of Toxicology *155*, 234–247.

Borchert, T., Hübscher, D., Guessoum, C.I., Lam, T.-D.D., Ghadri, J.R., Schellinger, I.N., Tiburcy, M., Liaw, N.Y., Li, Y., Haas, J., Sossalla, S., Huber, M.A., Cyganek, L., Jacobshagen, C., Dressel, R., Raaz, U., Nikolaev, V.O., Guan, K., Thiele, H., Meder, B., Wollnik, B., Zimmermann, W.-H., Lüscher, T.F., Hasenfuss, G., Templin, C., and Streckfuss-Bömeke, K. (2017). Catecholamine-Dependent β-Adrenergic Signaling in a Pluripotent Stem Cell Model of Takotsubo Cardiomyopathy. Journal of the American College of Cardiology *70*, 975–991.

Boriani, G., Biagini, E., Wahbi, K., and Duboc, D. (2015). Cardiac involvement in laminopathies. Orphanet Journal of Rare Diseases *10*, O25.

Brauch, K.M., Karst, M.L., Herron, K.J., Andrade, M. de, Pellikka, P.A., Rodeheffer, R.J., Michels, V.V., and Olson, T.M. (2009). Mutations in ribonucleic acid binding protein gene cause familial dilated cardiomyopathy. Journal of the American College of Cardiology *54*, 930–941.

Buckingham, M., Meilhac, S., and Zaffran, S. (2005). Building the mammalian heart from two sources of myocardial cells. Nature reviews. Genetics *6*, 826–835.

Burridge, P.W., Keller, G., Gold, J.D., and Wu, J.C. (2012). Production of de novo cardiomyocytes: human pluripotent stem cell differentiation and direct reprogramming. Cell stem cell *10*, 16–28.

7 REFERENCES

Burridge, P.W., and Zambidis, E.T. (2013). Highly efficient directed differentiation of human induced pluripotent stem cells into cardiomyocytes. Methods in molecular biology (Clifton, N.J.) *997*, 149–161.

Byrne, S.M., Mali, P., and Church, G.M. (2014). Genome editing in human stem cells. Methods in enzymology *546*, 119–138.

Cazalla, D., Zhu, J., Manche, L., Huber, E., Krainer, A.R., and Cáceres, J.F. (2002). Nuclear export and retention signals in the RS domain of SR proteins. Molecular and cellular biology *22*, 6871–6882.

Chelu, M.G., and Wehrens, X.H.T. (2007). Sarcoplasmic reticulum calcium leak and cardiac arrhythmias. Biochemical Society transactions *35*, 952–956.

Chen, G., Gulbranson, D.R., Hou, Z., Bolin, J.M., Ruotti, V., Probasco, M.D., Smuga-Otto, K., Howden, S.E., Diol, N.R., Propson, N.E., Wagner, R., Lee, G.O., Antosiewicz-Bourget, J., Teng, J.M.C., and Thomson, J.A. (2011). Chemically defined conditions for human iPSC derivation and culture. Nature methods *8*, 424–429.

Chen, H., Zhang, W., Li, D., Cordes, T.M., Mark Payne, R., and Shou, W. (2009). Analysis of ventricular hypertrabeculation and noncompaction using genetically engineered mouse models. Pediatric cardiology *30*, 626–634.

Chin, M.H., Mason, M.J., Xie, W., Volinia, S., Singer, M., Peterson, C., Ambartsumyan, G., Aimiuwu, O., Richter, L., Zhang, J., Khvorostov, I., Ott, V., Grunstein, M., Lavon, N., Benvenisty, N., Croce, C.M., Clark, A.T., Baxter, T., Pyle, A.D., Teitell, M.A., Pelegrini, M., Plath, K., and Lowry, W.E. (2009). Induced pluripotent stem cells and embryonic stem cells are distinguished by gene expression signatures. Cell stem cell *5*, 111–123.

Clark, K.A., McElhinny, A.S., Beckerle, M.C., and Gregorio, C.C. (2002). Striated muscle cytoarchitecture: an intricate web of form and function. Annual review of cell and developmental biology *18*, 637–706.

Clay, S., Alfakih, K., Radjenovic, A., Jones, T., Ridgway, J.P., and Sinvananthan, M.U. (2006). Normal range of human left ventricular volumes and mass using steady state free precession MRI in the radial long axis orientation. Magma (New York, N.Y.) *19*, 41–45.

Correia, C., Koshkin, A., Duarte, P., Hu, D., Teixeira, A., Domian, I., Serra, M., and Alves, P.M. (2017). Distinct carbon sources affect structural and functional maturation of cardiomyocytes derived from human pluripotent stem cells. Scientific reports *7*, 8590.

Cowan, J.R., Kinnamon, D.D., Morales, A., Salyer, L., Nickerson, D.A., and Hershberger, R.E. (2018). Multigenic Disease and Bilineal Inheritance in Dilated Cardiomyopathy Is Illustrated in Nonsegregating LMNA Pedigrees. Circulation. Genomic and precision medicine *11*, e002038.

Cyganek, L., Tiburcy, M., Sekeres, K., Gerstenberg, K., Bohnenberger, H., Lenz, C., Henze, S., Stauske, M., Salinas, G., Zimmermann, W.-H., Hasenfuss, G., and Guan, K. (2018). Deep phenotyping of human induced pluripotent stem cell-derived atrial and ventricular cardiomyocytes. JCI insight *3*.

Da Cruz, S., and Cleveland, D.W. (2011). Understanding the role of TDP-43 and FUS/TLS in ALS and beyond. Current opinion in neurobiology *21*, 904–919.

D'Amato, G., Luxán, G., and La Pompa, J.L. de (2016). Notch signalling in ventricular chamber development and cardiomyopathy. The FEBS journal *283*, 4223–4237.

Darche, F.F., Rivinius, R., Köllensperger, E., Leimer, U., Germann, G., Seckinger, A., Hose, D., Schröter, J., Bruehl, C., Draguhn, A., Gabriel, R., Schmidt, M., Koenen, M., Thomas, D., Katus, H.A., and Schweizer, P.A. (2019). Pacemaker cell characteristics of differentiated and HCN4-transduced human mesenchymal stem cells. Life sciences *232*, 116620.

Dauksaite, V., and Gotthardt, M. (2018). Molecular basis of titin exon exclusion by RBM20 and the novel titin splice regulator PTB4. Nucleic acids research *46*, 5227–5238.

Dedkova, E.N., and Blatter, L.A. (2013). Calcium signaling in cardiac mitochondria. Journal of molecular and cellular cardiology *58*, 125–133.

Deng, J., Shoemaker, R., Xie, B., Gore, A., LeProust, E.M., Antosiewicz-Bourget, J., Egli, D., Maherali, N., Park, I.-H., Yu, J., Daley, G.Q., Eggan, K., Hochedlinger, K., Thomson, J., Wang, W., Gao, Y., and Zhang, K. (2009). Targeted bisulfite sequencing reveals changes in DNA methylation associated with nuclear reprogramming. Nature biotechnology *27*, 353–360.

Dessauer, C.W., Watts, V.J., Ostrom, R.S., Conti, M., Dove, S., and Seifert, R. (2017). International Union of Basic and Clinical Pharmacology. CI. Structures and Small Molecule Modulators of Mammalian Adenylyl Cyclases. Pharmacological reviews *69*, 93–139.

Dominici, M., Le Blanc, K., Mueller, I., Slaper-Cortenbach, I., Marini, F., Krause, D., Deans, R., Keating, A., Prockop, D., and Horwitz, E. (2006). Minimal criteria for defining multipotent mesenchymal stromal cells. The International Society for Cellular Therapy position statement. Cytotherapy *8*, 315–317.

Dulhunty, A.F. (2006). Excitation-contraction coupling from the 1950s into the new millennium. Clinical and experimental pharmacology & physiology *33*, 763–772.

Eisner, D.A., Caldwell, J.L., Trafford, A.W., and Hutchings, D.C. (2020). The Control of Diastolic Calcium in the Heart: Basic Mechanisms and Functional Implications. Circulation research *126*, 395–412.

Eisner, D.A., Trafford, A.W., Dñaz, M.E., Overend, C.L., and O'Neill, S.C. (1998). The control of Ca release from the cardiac sarcoplasmic reticulum: regulation versus autoregulation. Cardiovascular research *38*, 589–604.

Elliott, P., Andersson, B., Arbustini, E., Bilinska, Z., Cecchi, F., Charron, P., Dubourg, O., Kühl, U., Maisch, B., McKenna, W.J., Monserrat, L., Pankuweit, S., Rapezzi, C., Seferovic, P., Tavazzi, L., and Keren, A. (2008). Classification of the cardiomyopathies: a position statement from the European Society Of Cardiology Working Group on Myocardial and Pericardial Diseases. European heart journal *29*, 270–276.

Evans, M.J., and Kaufman, M.H. (1981). Establishment in culture of pluripotential cells from mouse embryos. Nature *292*, 154–156.

7 REFERENCES

Faustino, N.A., and Cooper, T.A. (2003). Pre-mRNA splicing and human disease. Genes & development *17*, 419–437.

Filippello, A., Lorenzi, P., Bergamo, E., and Romanelli, M.G. (2013). Identification of nuclear retention domains in the RBM20 protein. FEBS letters *587*, 2989–2995.

Finsterer, J. (2019). Barth syndrome: mechanisms and management. The application of clinical genetics *12*, 95–106.

Freiburg, A., Trombitas, K., Hell, W., Cazorla, O., Fougerousse, F., Centner, T., Kolmerer, B., Witt, C., Beckmann, J.S., Gregorio, C.C., Granzier, H., and Labeit, S. (2000). Series of exon-skipping events in the elastic spring region of titin as the structural basis for myofibrillar elastic diversity. Circulation research *86*, 1114–1121.

Frese, K.S., Meder, B., Keller, A., Just, S., Haas, J., Vogel, B., Fischer, S., Backes, C., Matzas, M., Köhler, D., Benes, V., Katus, H.A., and Rottbauer, W. (2015). RNA splicing regulated by RBFOX1 is essential for cardiac function in zebrafish. Journal of Cell Science *128*, 3030–3040.

Fusaki, N., Ban, H., Nishiyama, A., Saeki, K., and Hasegawa, M. (2009). Efficient induction of transgene-free human pluripotent stem cells using a vector based on Sendai virus, an RNA virus that does not integrate into the host genome. Proceedings of the Japan Academy. Series B, Physical and Biological Sciences *85*, 348–362.

Gautel, M., and Djinović-Carugo, K. (2016). The sarcomeric cytoskeleton: from molecules to motion. The Journal of experimental biology *219*, 135–145.

George, C.H., Jundi, H., Walters, N., Thomas, N.L., West, R.R., and Lai, F.A. (2006). Arrhythmogenic mutation-linked defects in ryanodine receptor autoregulation reveal a novel mechanism of Ca2+ release channel dysfunction. Circulation research *98*, 88–97.

George, C.H., Rogers, S.A., Bertrand, B.M.A., Tunwell, R.E.A., Thomas, N.L., Steele, D.S., Cox, E.V., Pepper, C., Hazeel, C.J., Claycomb, W.C., and Lai, F.A. (2007). Alternative splicing of ryanodine receptors modulates cardiomyocyte Ca2+ signaling and susceptibility to apoptosis. Circulation research *100*, 874–883.

George, R.M., and Firulli, A.B. (2019). Hand Factors in Cardiac Development. Anatomical record (Hoboken, N.J. 2007) *302*, 101–107.

Geuens, T., Bouhy, D., and Timmerman, V. (2016). The hnRNP family: insights into their role in health and disease. Human genetics *135*, 851–867.

Grant, A.O. (2009). Cardiac ion channels. Circulation. Arrhythmia and electrophysiology *2*, 185–194.

Gray, C.B.B., and Heller Brown, J. (2014). CaMKIIdelta subtypes: localization and function. Frontiers in Pharmacology *5*, 15.

Grimm, M., and Brown, J.H. (2010). Beta-adrenergic receptor signaling in the heart: role of CaMKII. Journal of molecular and cellular cardiology *48*, 322–330.

Grissa, I., Vergnaud, G., and Pourcel, C. (2007). The CRISPRdb database and tools to display CRISPRs and to generate dictionaries of spacers and repeats. BMC bioinformatics *8*, 172.

Guan, K., Nayernia, K., Maier, L.S., Wagner, S., Dressel, R., Lee, J.H., Nolte, J., Wolf, F., Li, M., Engel, W., and Hasenfuss, G. (2006). Pluripotency of spermatogonial stem cells from adult mouse testis. Nature *440*, 1199–1203.

Guenther, M.G., Frampton, G.M., Soldner, F., Hockemeyer, D., Mitalipova, M., Jaenisch, R., and Young, R.A. (2010). Chromatin structure and gene expression programs of human embryonic and induced pluripotent stem cells. Cell stem cell *7*, 249–257.

Guo, W., Schafer, S., Greaser, M.L., Radke, M.H., Liss, M., Govindarajan, T., Maatz, H., Schulz, H., Li, S., Parrish, A.M., Dauksaite, V., Vakeel, P., Klaassen, S., Gerull, B., Thierfelder, L., Regitz-Zagrosek, V., Hacker, T.A., Saupe, K.W., Dec, G.W., Ellinor, P.T., MacRae, C.A., Spallek, B., Fischer, R., Perrot, A., Özcelik, C., Saar, K., Hubner, N., and Gotthardt, M. (2012). RBM20, a gene for hereditary cardiomyopathy, regulates titin splicing. Nature medicine *18*, 766–773.

Haft, D.H., Selengut, J., Mongodin, E.F., and Nelson, K.E. (2005). A guild of 45 CRISPR-associated (Cas) protein families and multiple CRISPR/Cas subtypes exist in prokaryotic genomes. PLoS computational biology *1*, e60.

Hall, P.A., and Watt, F.M. (1989). Stem cells: the generation and maintenance of cellular diversity. Development (Cambridge, England) *106*, 619–633.

Hancox, J.C., James, A.F., Walsh, M.A., and Stuart, A.G. (2017). Triadin mutations - a cause of ventricular arrhythmias in children and young adults. Journal of Congenital Cardiology *1*.

Hastings, M.L., and Krainer, A.R. (2001). Pre-mRNA splicing in the new millennium. Current Opinion in Cell Biology *13*, 302–309.

Herman, C.A., and Sandoval, E.J. (1983). Catecholamine effects on blood pressure and heart rate in the American bullfrog, Rana catesbeiana. General and Comparative Endocrinology *52*, 142–148.

Herman, D.S., Lam, L., Taylor, M.R.G., Wang, L., Teekakirikul, P., Christodoulou, D., Conner, L., DePalma, S.R., McDonough, B., Sparks, E., Teodorescu, D.L., Cirino, A.L., Banner, N.R., Pennell, D.J., Graw, S., Merlo, M., Di Lenarda, A., Sinagra, G., Bos, J.M., Ackerman, M.J., Mitchell, R.N., Murry, C.E., Lakdawala, N.K., Ho, C.Y., Barton, P.J.R., Cook, S.A., Mestroni, L., Seidman, J.G., and Seidman, C.E. (2012). Truncations of titin causing dilated cardiomyopathy. The New England journal of medicine *366*, 619–628.

Hershberger, R.E., Hedges, D.J., and Morales, A. (2013). Dilated cardiomyopathy: the complexity of a diverse genetic architecture. Nature reviews. Cardiology *10*, 531–547.

Hey, T.M., Rasmussen, T.B., Madsen, T., Aagaard, M.M., Harbo, M., Mølgaard, H., Møller, J.E., Eiskjær, H., and Mogensen, J. (2019). Pathogenic RBM20-Variants Are Associated With a Severe Disease Expression in Male Patients With Dilated Cardiomyopathy. Circulation. Heart failure *12*, e005700.

Horii, T., Tamura, D., Morita, S., Kimura, M., and Hatada, I. (2013). Generation of an ICF syndrome model by efficient genome editing of human induced pluripotent stem cells using the CRISPR system. International journal of molecular sciences *14*, 19774–19781.

Horikoshi, Y., Yan, Y., Terashvili, M., Wells, C., Horikoshi, H., Fujita, S., Bosnjak, Z.J., and Bai, X. (2019). Fatty Acid-Treated Induced Pluripotent Stem Cell-Derived Human Cardiomyocytes Exhibit Adult Cardiomyocyte-Like Energy Metabolism Phenotypes. Cells *8*.

Horváth, A., Lemoine, M.D., Löser, A., Mannhardt, I., Flenner, F., Uzun, A.U., Neuber, C., Breckwoldt, K., Hansen, A., Girdauskas, E., Reichenspurner, H., Willems, S., Jost, N., Wettwer, E., Eschenhagen, T., and Christ, T. (2018). Low Resting Membrane Potential and Low Inward Rectifier Potassium Currents Are Not Inherent Features of hiPSC-Derived Cardiomyocytes. Stem cell reports *10*, 822–833.

Hu, K., Yu, J., Suknuntha, K., Tian, S., Montgomery, K., Choi, K.-D., Stewart, R., Thomson, J.A., and Slukvin, I.I. (2011). Efficient generation of transgene-free induced pluripotent stem cells from normal and neoplastic bone marrow and cord blood mononuclear cells. Blood *117*, e109-19.

Hubé, F., and Francastel, C. (2015). Mammalian introns: when the junk generates molecular diversity. International journal of molecular sciences *16*, 4429–4452.

Hübscher, D., Rebs, S., Haupt, L., Borchert, T., Guessoum, C.I., Treu, F., Köhne, S., Maus, A., Hambrecht, M., Sossalla, S., Dressel, R., Uy, A., Jakob, M., Hasenfuss, G., and Streckfuss-Bömeke, K. (2019). A High-Throughput Method as a Diagnostic Tool for HIV Detection in Patient-Specific Induced Pluripotent Stem Cells Generated by Different Reprogramming Methods. Stem cells international *2019*, 2181437.

Huxley, H.E. (1969). The mechanism of muscular contraction. Science (New York, N.Y.) *164*, 1356–1365.

Ishino, Y., Shinagawa, H., Makino, K., Amemura, M., and Nakata, A. (1987). Nucleotide sequence of the iap gene, responsible for alkaline phosphatase isozyme conversion in Escherichia coli, and identification of the gene product. Journal of bacteriology *169*, 5429–5433.

Ito, Y., Kawamorita, M., Yamabe, T., Kiyono, T., and Miyamoto, K. (2007). Chemically fixed nurse cells for culturing murine or primate embryonic stem cells. Journal of bioscience and bioengineering *103*, 113–121.

Jansen, R., van Embden, J.D.A., Gaastra, W., and Schouls, L.M. (2002). Identification of genes that are associated with DNA repeats in prokaryotes. Molecular microbiology *43*, 1565–1575.

Jenni, R., Oechslin, E.N., and van der Loo, B. (2007). Isolated ventricular non-compaction of the myocardium in adults. Heart (British Cardiac Society) *93*, 11–15.

Jinek, M., Chylinski, K., Fonfara, I., Hauer, M., Doudna, J.A., and Charpentier, E. (2012). A programmable dual-RNA-guided DNA endonuclease in adaptive bacterial immunity. Science (New York, N.Y.) *337*, 816–821.

Juan-Mateu, J., González-Quereda, L., Rodríguez, M.J., Verdura, E., Lázaro, K., Jou, C., Nascimento, A., Jiménez-Mallebrera, C., Colomer, J., Monges, S., Lubieniecki, F., Foncuberta, M.E., Pascual-Pascual,

S.I., Molano, J., Baiget, M., and Gallano, P. (2013). Interplay between DMD Point Mutations and Splicing Signals in Dystrophinopathy Phenotypes. PloS one *8*.

Kamakura, T., Makiyama, T., Sasaki, K., Yoshida, Y., Wuriyanghai, Y., Chen, J., Hattori, T., Ohno, S., Kita, T., Horie, M., Yamanaka, S., and Kimura, T. (2013). Ultrastructural maturation of human-induced pluripotent stem cell-derived cardiomyocytes in a long-term culture. Circulation journal official journal of the Japanese Circulation Society *77*, 1307–1314.

Kamisago, M., Sharma, S.D., DePalma, S.R., Solomon, S., Sharma, P., McDonough, B., Smoot, L., Mullen, M.P., Woolf, P.K., Wigle, E.D., Seidman, J.G., and Seidman, C.E. (2000). Mutations in sarcomere protein genes as a cause of dilated cardiomyopathy. The New England journal of medicine *343*, 1688–1696.

Kampmann, M. (2018). CRISPRi and CRISPRa Screens in Mammalian Cells for Precision Biology and Medicine. ACS chemical biology *13*, 406–416.

Kapa, S., Tester, D.J., Salisbury, B.A., Harris-Kerr, C., Pungliya, M.S., Alders, M., Wilde, A.A.M., and Ackerman, M.J. (2009). Genetic testing for long-QT syndrome: distinguishing pathogenic mutations from benign variants. Circulation *120*, 1752–1760.

Karakikes, I., Ameen, M., Termglinchan, V., and Wu, J.C. (2015). Human induced pluripotent stem cell-derived cardiomyocytes: insights into molecular, cellular, and functional phenotypes. Circulation research *117*, 80–88.

Karakikes, I., Termglinchan, V., and Wu, J.C. (2014). Human-induced pluripotent stem cell models of inherited cardiomyopathies. Current opinion in cardiology *29*, 214–219.

Kelly, S.J. (1977). Studies of the developmental potential of 4- and 8-cell stage mouse blastomeres. The Journal of experimental zoology *200*, 365–376.

Khan, M.A.F., Reckman, Y.J., Aufiero, S., van den Hoogenhof, M.M.G., van der Made, I., Beqqali, A., Koolbergen, D.R., Rasmussen, T.B., van der Velden, J., Creemers, E.E., and Pinto, Y.M. (2016). RBM20 Regulates Circular RNA Production From the Titin Gene. Circulation research *119*, 996–1003.

Kim, D., Kim, C.-H., Moon, J.-I., Chung, Y.-G., Chang, M.-Y., Han, B.-S., Ko, S., Yang, E., Cha, K.Y., Lanza, R., and Kim, K.-S. (2009). Generation of human induced pluripotent stem cells by direct delivery of reprogramming proteins. Cell stem cell *4*, 472–476.

Kim, R.Y., Robertson, E.J., and Solloway, M.J. (2001). Bmp6 and Bmp7 are required for cushion formation and septation in the developing mouse heart. Developmental biology *235*, 449–466.

Kirchhefer, U. (1999). Activity of cAMP-dependent protein kinase and Ca2+/calmodulin-dependent protein kinase in failing and nonfailing human hearts. Cardiovascular research *42*, 254–261.

Kodo, K., Ong, S.-G., Jahanbani, F., Termglinchan, V., Hirono, K., InanlooRahatloo, K., Ebert, A.D., Shukla, P., Abilez, O.J., Churko, J.M., Karakikes, I., Jung, G., Ichida, F., Wu, S.M., Snyder, M.P., Bernstein, D., and Wu, J.C. (2016). iPSC-derived cardiomyocytes reveal abnormal TGF-β signalling in left ventricular non-compaction cardiomyopathy. Nature cell biology *18*, 1031–1042.

7 REFERENCES

Kolanowski, T.J., Antos, C.L., and Guan, K. (2017). Making human cardiomyocytes up to date: Derivation, maturation state and perspectives. International Journal of Cardiology *241*, 379–386.

Kranias, E.G., and Bers, D.M. (2007). Calcium and cardiomyopathies. Sub-cellular biochemistry *45*, 523–537.

Krawczak, M., Thomas, N.S.T., Hundrieser, B., Mort, M., Wittig, M., Hampe, J., and Cooper, D.N. (2007). Single base-pair substitutions in exon-intron junctions of human genes: nature, distribution, and consequences for mRNA splicing. Human mutation *28*, 150–158.

Krcmery, J., Camarata, T., Kulisz, A., and Simon, H.-G. (2010). Nucleocytoplasmic functions of the PDZ-LIM protein family: new insights into organ development. BioEssays news and reviews in molecular, cellular and developmental biology *32*, 100–108.

Krikler, D.M. (1986). Verapamil in arrhythmia. British journal of clinical pharmacology *21 Suppl 2*, 183S-189S.

LaBarge, W., Mattappally, S., Kannappan, R., Fast, V.G., Pretorius, D., Berry, J.L., and Zhang, J. (2019). Maturation of three-dimensional, hiPSC-derived cardiomyocyte spheroids utilizing cyclic, uniaxial stretch and electrical stimulation. PloS one *14*, e0219442.

Lan, F., Lee, A.S., Liang, P., Sanchez-Freire, V., Nguyen, P.K., Wang, L., Han, L., Yen, M., Wang, Y., Sun, N., Abilez, O.J., Hu, S., Ebert, A.D., Navarrete, E.G., Simmons, C.S., Wheeler, M., Pruitt, B., Lewis, R., Yamaguchi, Y., Ashley, E.A., Bers, D.M., Robbins, R.C., Longaker, M.T., and Wu, J.C. (2013). Abnormal calcium handling properties underlie familial hypertrophic cardiomyopathy pathology in patient-specific induced pluripotent stem cells. Cell stem cell *12*, 101–113.

Landstrom, A.P., Dobrev, D., and Wehrens, X.H.T. (2017). Calcium Signaling and Cardiac Arrhythmias. Circulation research *120*, 1969–1993.

Lee, Y., and Rio, D.C. (2015). Mechanisms and Regulation of Alternative Pre-mRNA Splicing. Annual review of biochemistry *84*, 291–323.

Lehman, W., Craig, R., and Vibert, P. (1994). Ca(2+)-induced tropomyosin movement in Limulus thin filaments revealed by three-dimensional reconstruction. Nature *368*, 65–67.

Lemoine, M.D., Mannhardt, I., Breckwoldt, K., Prondzynski, M., Flenner, F., Ulmer, B., Hirt, M.N., Neuber, C., Horváth, A., Kloth, B., Reichenspurner, H., Willems, S., Hansen, A., Eschenhagen, T., and Christ, T. (2017). Human iPSC-derived cardiomyocytes cultured in 3D engineered heart tissue show physiological upstroke velocity and sodium current density. Scientific reports *7*, 5464.

Lewandowski, J., Rozwadowska, N., Kolanowski, T.J., Malcher, A., Zimna, A., Rugowska, A., Fiedorowicz, K., Łabędź, W., Kubaszewski, Ł., Chojnacka, K., Bednarek-Rajewska, K., Majewski, P., and Kurpisz, M. (2018). The impact of in vitro cell culture duration on the maturation of human cardiomyocytes derived from induced pluripotent stem cells of myogenic origin. Cell transplantation *27*, 1047–1067.

Li, D., Morales, A., Gonzalez-Quintana, J., Norton, N., Siegfried, J.D., Hofmeyer, M., and Hershberger, R.E. (2010). Identification of novel mutations in RBM20 in patients with dilated cardiomyopathy. Clinical and translational science 3, 90–97.

Li, D., Zhou, H., and Zeng, X. (2019a). Battling CRISPR-Cas9 off-target genome editing. Cell biology and toxicology 35, 403–406.

Li, J., Wen, A.M., Potla, R., Benshirim, E., Seebarran, A., Benz, M.A., Henry, O.Y.F., Matthews, B.D., Prantil-Baun, R., Gilpin, S.E., Levy, O., and Ingber, D.E. (2019b). AAV-mediated gene therapy targeting TRPV4 mechanotransduction for inhibition of pulmonary vascular leakage. APL bioengineering 3, 46103.

Lian, X., Zhang, J., Azarin, S.M., Zhu, K., Hazeltine, L.B., Bao, X., Hsiao, C., Kamp, T.J., and Palecek, S.P. (2013). Directed cardiomyocyte differentiation from human pluripotent stem cells by modulating Wnt/β-catenin signaling under fully defined conditions. Nature protocols 8, 162–175.

Liang, X., Potter, J., Kumar, S., Zou, Y., Quintanilla, R., Sridharan, M., Carte, J., Chen, W., Roark, N., Ranganathan, S., Ravinder, N., and Chesnut, J.D. (2015). Rapid and highly efficient mammalian cell engineering via Cas9 protein transfection. Journal of biotechnology 208, 44–53.

Lim, K.H., Ferraris, L., Filloux, M.E., Raphael, B.J., and Fairbrother, W.G. (2011). Using positional distribution to identify splicing elements and predict pre-mRNA processing defects in human genes. Proceedings of the National Academy of Sciences of the United States of America 108, 11093–11098.

Lino, C.A., Harper, J.C., Carney, J.P., and Timlin, J.A. (2018). Delivering CRISPR: a review of the challenges and approaches. Drug delivery 25, 1234–1257.

Lou, Q., Janardhan, A., and Efimov, I.R. (2012). Remodeling of calcium handling in human heart failure. Advances in experimental medicine and biology 740, 1145–1174.

Lozano, R., Naghavi, M., Foreman, K., Lim, S., Shibuya, K., Aboyans, V., Abraham, J., Adair, T., Aggarwal, R., Ahn, S.Y., AlMazroa, M.A., Alvarado, M., Anderson, H.R., Anderson, L.M., Andrews, K.G., Atkinson, C., Baddour, L.M., Barker-Collo, S., Bartels, D.H., Bell, M.L., Benjamin, E.J., Bennett, D., Bhalla, K., Bikbov, B., Abdulhak, A.B., Birbeck, G., Blyth, F., Bolliger, I., Boufous, S., Bucello, C., Burch, M., Burney, P., Carapetis, J., Chen, H., Chou, D., Chugh, S.S., Coffeng, L.E., Colan, S.D., Colquhoun, S., Colson, K.E., Condon, J., Connor, M.D., Cooper, L.T., Corriere, M., Cortinovis, M., Vaccaro, K.C. de, Couser, W., Cowie, B.C., Criqui, M.H., Cross, M., Dabhadkar, K.C., Dahodwala, N., Leo, D. de, Degenhardt, L., Delossantos, A., Denenberg, J., Des Jarlais, D.C., Dharmaratne, S.D., Dorsey, E.R., Driscoll, T., Duber, H., Ebel, B., Erwin, P.J., Espindola, P., Ezzati, M., Feigin, V., Flaxman, A.D., Forouzanfar, M.H., Fowkes, F.G.R., Franklin, R., Fransen, M., Freeman, M.K., Gabriel, S.E., Gakidou, E., Gaspari, F., Gillum, R.F., Gonzalez-Medina, D., Halasa, Y.A., Haring, D., Harrison, J.E., Havmoeller, R., Hay, R.J., Hoen, B., Hotez, P.J., Hoy, D., Jacobsen, K.H., James, S.L., Jasrasaria, R., Jayaraman, S., Johns, N., Karthikeyan, G., Kassebaum, N., Keren, A., Khoo, J.-P., Knowlton, L.M., Kobusingye, O., Koranteng, A., Krishnamurthi, R., Lipnick, M., Lipshultz, S.E., Ohno, S.L., Mabweijano, J., MacIntyre, M.F., Mallinger, L., March, L., Marks, G.B., Marks, R., Matsumori, A., Matzopoulos, R., Mayosi, B.M., McAnulty, J.H., McDermott, M.M., McGrath, J., Memish, Z.A., Mensah, G.A., Merriman, T.R., Michaud, C., Miller, M., Miller, T.R., Mock, C., Mocumbi, A.O., Mokdad, A.A., Moran, A., Mulholland,

K., Nair, M.N., Naldi, L., Narayan, K.M.V., Nasseri, K., Norman, P., O'Donnell, M., Omer, S.B., Ortblad, K., Osborne, R., Ozgediz, D., Pahari, B., Pandian, J.D., Rivero, A.P., Padilla, R.P., Perez-Ruiz, F., Perico, N., Phillips, D., Pierce, K., Pope, C.A., Porrini, E., Pourmalek, F., Raju, M., Ranganathan, D., Rehm, J.T., Rein, D.B., Remuzzi, G., Rivara, F.P., Roberts, T., León, F.R. de, Rosenfeld, L.C., Rushton, L., Sacco, R.L., Salomon, J.A., Sampson, U., Sanman, E., Schwebel, D.C., Segui-Gomez, M., Shepard, D.S., Singh, D., Singleton, J., Sliwa, K., Smith, E., Steer, A., Taylor, J.A., Thomas, B., Tleyjeh, I.M., Towbin, J.A., Truelsen, T., Undurraga, E.A., Venketasubramanian, N., Vijayakumar, L., Vos, T., Wagner, G.R., Wang, M., Wang, W., Watt, K., Weinstock, M.A., Weintraub, R., Wilkinson, J.D., Woolf, A.D., Wulf, S., Yeh, P.-H., Yip, P., Zabetian, A., Zheng, Z.-J., Lopez, A.D., and Murray, C.J.L. (2012). Global and regional mortality from 235 causes of death for 20 age groups in 1990 and 2010: a systematic analysis for the Global Burden of Disease Study 2010. The Lancet *380*, 2095–2128.

Lundy, S.D., Zhu, W.-Z., Regnier, M., and Laflamme, M.A. (2013). Structural and functional maturation of cardiomyocytes derived from human pluripotent stem cells. Stem cells and development *22*, 1991–2002.

Maatz, H., Jens, M., Liss, M., Schafer, S., Heinig, M., Kirchner, M., Adami, E., Rintisch, C., Dauksaite, V., Radke, M.H., Selbach, M., Barton, P.J.R., Cook, S.A., Rajewsky, N., Gotthardt, M., Landthaler, M., and Hubner, N. (2014). RNA-binding protein RBM20 represses splicing to orchestrate cardiac pre-mRNA processing. The Journal of clinical investigation *124*, 3419–3430.

Machiraju, P., and Greenway, S.C. (2019). Current methods for the maturation of induced pluripotent stem cell-derived cardiomyocytes. World Journal of Stem Cells *11*, 33–43.

Maeder, M.L., Stefanidakis, M., Wilson, C.J., Baral, R., Barrera, L.A., Bounoutas, G.S., Bumcrot, D., Chao, H., Ciulla, D.M., DaSilva, J.A., Dass, A., Dhanapal, V., Fennell, T.J., Friedland, A.E., Giannoukos, G., Gloskowski, S.W., Glucksmann, A., Gotta, G.M., Jayaram, H., Haskett, S.J., Hopkins, B., Horng, J.E., Joshi, S., Marco, E., Mepani, R., Reyon, D., Ta, T., Tabbaa, D.G., Samuelsson, S.J., Shen, S., Skor, M.N., Stetkiewicz, P., Wang, T., Yudkoff, C., Myer, V.E., Albright, C.F., and Jiang, H. (2019). Development of a gene-editing approach to restore vision loss in Leber congenital amaurosis type 10. Nature medicine *25*, 229–233.

Makarova, K.S., Wolf, Y.I., Alkhnbashi, O.S., Costa, F., Shah, S.A., Saunders, S.J., Barrangou, R., Brouns, S.J.J., Charpentier, E., Haft, D.H., Horvath, P., Moineau, S., Mojica, F.J.M., Terns, R.M., Terns, M.P., White, M.F., Yakunin, A.F., Garrett, R.A., van der Oost, J., Backofen, R., and Koonin, E.V. (2015). An updated evolutionary classification of CRISPR-Cas systems. Nature reviews. Microbiology *13*, 722–736.

Malkani, R., D'Souza, I., Gwinn-Hardy, K., Schellenberg, G.D., Hardy, J., and Momeni, P. (2006). A MAPT mutation in a regulatory element upstream of exon 10 causes frontotemporal dementia. Neurobiology of disease *22*, 401–403.

Marks, A.R. (2013). Calcium cycling proteins and heart failure: mechanisms and therapeutics. The Journal of clinical investigation *123*, 46–52.

Maron, B.J., Towbin, J.A., Thiene, G., Antzelevitch, C., Corrado, D., Arnett, D., Moss, A.J., Seidman, C.E., and Young, J.B. (2006). Contemporary definitions and classification of the cardiomyopathies: an

7 REFERENCES

American Heart Association Scientific Statement from the Council on Clinical Cardiology, Heart Failure and Transplantation Committee; Quality of Care and Outcomes Research and Functional Genomics and Translational Biology Interdisciplinary Working Groups; and Council on Epidemiology and Prevention. Circulation *113*, 1807–1816.

Mattiazzi, A., and Kranias, E.G. (2014). The role of CaMKII regulation of phospholamban activity in heart disease. Frontiers in Pharmacology *5*, 5.

McKenna, W.J., Maron, B.J., and Thiene, G. (2017). Classification, Epidemiology, and Global Burden of Cardiomyopathies. Circulation research *121*, 722–730.

McNair, W.P., Ku, L., Taylor, M.R.G., Fain, P.R., Dao, D., Wolfel, E., and Mestroni, L. (2004). SCN5A mutation associated with dilated cardiomyopathy, conduction disorder, and arrhythmia. Circulation *110*, 2163–2167.

McNally, E.M., and Mestroni, L. (2017). Dilated Cardiomyopathy: Genetic Determinants and Mechanisms. Circulation research *121*, 731–748.

Miao, L., Li, J., Li, J., Tian, X., Lu, Y., Hu, S., Shieh, D., Kanai, R., Zhou, B.-Y., Zhou, B., Liu, J., Firulli, A.B., Martin, J.F., Singer, H., Xin, H., and Wu, M. (2018). Notch signaling regulates Hey2 expression in a spatiotemporal dependent manner during cardiac morphogenesis and trabecular specification. Scientific reports *8*, 2678.

Miszalski-Jamka, K., Jefferies, J.L., Mazur, W., Głowacki, J., Hu, J., Lazar, M., Gibbs, R.A., Liczko, J., Kłyś, J., Venner, E., Muzny, D.M., Rycaj, J., Białkowski, J., Kluczewska, E., Kalarus, Z., Jhangiani, S., Al-Khalidi, H., Kukulski, T., Lupski, J.R., Craigen, W.J., and Bainbridge, M.N. (2017). Novel Genetic Triggers and Genotype-Phenotype Correlations in Patients With Left Ventricular Noncompaction. Circulation. Cardiovascular genetics *10*.

Moorman, A., Webb, S., Brown, N.A., Lamers, W., and Anderson, R.H. (2003). Development of the heart: (1) formation of the cardiac chambers and arterial trunks. Heart (British Cardiac Society) *89*, 806–814.

Moretti, A., Fonteyne, L., Giesert, F., Hoppmann, P., Meier, A.B., Bozoglu, T., Baehr, A., Schneider, C.M., Sinnecker, D., Klett, K., Fröhlich, T., Rahman, F.A., Haufe, T., Sun, S., Jurisch, V., Kessler, B., Hinkel, R., Dirschinger, R., Martens, E., Jilek, C., Graf, A., Krebs, S., Santamaria, G., Kurome, M., Zakhartchenko, V., Campbell, B., Voelse, K., Wolf, A., Ziegler, T., Reichert, S., Lee, S., Flenkenthaler, F., Dorn, T., Jeremias, I., Blum, H., Dendorfer, A., Schnieke, A., Krause, S., Walter, M.C., Klymiuk, N., Laugwitz, K.L., Wolf, E., Wurst, W., and Kupatt, C. (2020). Somatic gene editing ameliorates skeletal and cardiac muscle failure in pig and human models of Duchenne muscular dystrophy. Nature medicine *26*, 207–214.

Murayama, R., Kimura-Asami, M., Togo-Ohno, M., Yamasaki-Kato, Y., Naruse, T.K., Yamamoto, T., Hayashi, T., Ai, T., Spoonamore, K.G., Kovacs, R.J., Vatta, M., Iizuka, M., Saito, M., Wani, S., Hiraoka, Y., Kimura, A., and Kuroyanagi, H. (2018). Phosphorylation of the RSRSP stretch is critical for splicing regulation by RNA-Binding Motif Protein 20 (RBM20) through nuclear localization. Scientific reports *8*, 8970.

Nagueh, S.F., Shah, G., Wu, Y., Torre-Amione, G., King, N.M.P., Lahmers, S., Witt, C.C., Becker, K., Labeit, S., and Granzier, H.L. (2004). Altered titin expression, myocardial stiffness, and left ventricular function in patients with dilated cardiomyopathy. Circulation *110*, 155–162.

Najafi, A., Sequeira, V., Kuster, D.W.D., and van der Velden, J. (2016). β-adrenergic receptor signalling and its functional consequences in the diseased heart. European journal of clinical investigation *46*, 362–374.

Natividad-Diaz, S.L., Browne, S., Jha, A.K., Ma, Z., Hossainy, S., Kurokawa, Y.K., George, S.C., and Healy, K.E. (2019). A combined hiPSC-derived endothelial cell and in vitro microfluidic platform for assessing biomaterial-based angiogenesis. Biomaterials *194*, 73–83.

Neagoe, C., Kulke, M., del Monte, F., Gwathmey, J.K., Tombe, P.P. de, Hajjar, R.J., and Linke, W.A. (2002). Titin isoform switch in ischemic human heart disease. Circulation *106*, 1333–1341.

Ogawa, M. (1993). Differentiation and proliferation of hematopoietic stem cells. Blood *81*, 2844–2853.

Ohnuki, M., and Takahashi, K. (2015). Present and future challenges of induced pluripotent stem cells. Philosophical transactions of the Royal Society of London. Series B, Biological sciences *370*, 20140367.

Okamoto, S., Amaishi, Y., Maki, I., Enoki, T., and Mineno, J. (2019). Highly efficient genome editing for single-base substitutions using optimized ssODNs with Cas9-RNPs. Scientific reports *9*, 4811.

Okita, K., Matsumura, Y., Sato, Y., Okada, A., Morizane, A., Okamoto, S., Hong, H., Nakagawa, M., Tanabe, K., Tezuka, K.-i., Shibata, T., Kunisada, T., Takahashi, M., Takahashi, J., Saji, H., and Yamanaka, S. (2011). A more efficient method to generate integration-free human iPS cells. Nature methods *8*, 409–412.

Oshiro, C., Thorn, C.F., Roden, D.M., Klein, T.E., and Altman, R.B. (2010). KCNH2 pharmacogenomics summary. Pharmacogenetics and genomics *20*, 775–777.

Pan, Q., Shai, O., Lee, L.J., Frey, B.J., and Blencowe, B.J. (2008). Deep surveying of alternative splicing complexity in the human transcriptome by high-throughput sequencing. Nature genetics *40*, 1413–1415.

Paquet, D., Kwart, D., Chen, A., Sproul, A., Jacob, S., Teo, S., Olsen, K.M., Gregg, A., Noggle, S., and Tessier-Lavigne, M. (2016). Efficient introduction of specific homozygous and heterozygous mutations using CRISPR/Cas9. Nature *533*, 125–129.

Park, I.-H., Zhao, R., West, J.A., Yabuuchi, A., Huo, H., Ince, T.A., Lerou, P.H., Lensch, M.W., and Daley, G.Q. (2008). Reprogramming of human somatic cells to pluripotency with defined factors. Nature *451*, 141–146.

Peretto, G., Di Resta, C., Perversi, J., Forleo, C., Maggi, L., Politano, L., Barison, A., Previtali, S.C., Carboni, N., Brun, F., Pegoraro, E., D'Amico, A., Rodolico, C., Magri, F., Manzi, R.C., Palladino, A., Isola, F., Gigli, L., Mongini, T.E., Semplicini, C., Calore, C., Ricci, G., Comi, G.P., Ruggiero, L., Bertini, E., Bonomo, P., Nigro, G., Resta, N., Emdin, M., Favale, S., Siciliano, G., Santoro, L., Sinagra, G.,

Limongelli, G., Ambrosi, A., Ferrari, M., Golzio, P.G., Della Bella, P., Benedetti, S., and Sala, S. (2019). Cardiac and Neuromuscular Features of Patients With LMNA-Related Cardiomyopathy. Annals of internal medicine *171*, 458–463.

Pfeiffer, P., and Vielmetter, W. (1988). Joining of nonhomologous DNA double strand breaks in vitro. Nucleic acids research *16*, 907–924.

Phillips, B.T., Gassei, K., and Orwig, K.E. (2010). Spermatogonial stem cell regulation and spermatogenesis. Philosophical transactions of the Royal Society of London. Series B, Biological sciences *365*, 1663–1678.

Pickar-Oliver, A., and Gersbach, C.A. (2019). The next generation of CRISPR-Cas technologies and applications. Nature reviews. Molecular cell biology *20*, 490–507.

Priest, B.T., and McDermott, J.S. (2015). Cardiac ion channels. Channels (Austin, Tex.) *9*, 352–359.

Prondzynski, M., Lemoine, M.D., Zech, A.T., Horváth, A., Di Mauro, V., Koivumäki, J.T., Kresin, N., Busch, J., Krause, T., Krämer, E., Schlossarek, S., Spohn, M., Friedrich, F.W., Münch, J., Laufer, S.D., Redwood, C., Volk, A.E., Hansen, A., Mearini, G., Catalucci, D., Meyer, C., Christ, T., Patten, M., Eschenhagen, T., and Carrier, L. (2019). Disease modeling of a mutation in α-actinin 2 guides clinical therapy in hypertrophic cardiomyopathy. EMBO molecular medicine *11*, e11115.

Qiao, X.-H., Wang, F., Zhang, X.-L., Huang, R.-T., Xue, S., Wang, J., Qiu, X.-B., Liu, X.-Y., and Yang, Y.-Q. (2017). MEF2C loss-of-function mutation contributes to congenital heart defects. International journal of medical sciences *14*, 1143–1153.

Rappsilber, J., Ryder, U., Lamond, A.I., and Mann, M. (2002). Large-scale proteomic analysis of the human spliceosome. Genome research *12*, 1231–1245.

Rebs, S., Sedaghat-Hamedani, F., Kayvanpour, E., Meder, B., and Streckfuss-Bömeke, K. (2020). Generation of pluripotent stem cell lines and CRISPR/Cas9 modified isogenic controls from a patient with dilated cardiomyopathy harboring a RBM20 p.R634W mutation. Stem cell research *47*, 101901.

Regitz-Zagrosek, V., Petrov, G., Lehmkuhl, E., Smits, J.M., Babitsch, B., Brunhuber, C., Jurmann, B., Stein, J., Schubert, C., Merz, N.B., Lehmkuhl, H.B., and Hetzer, R. (2010). Heart transplantation in women with dilated cardiomyopathy. Transplantation *89*, 236–244.

Richter, C., Chang, J.T., and Fineran, P.C. (2012). Function and regulation of clustered regularly interspaced short palindromic repeats (CRISPR) / CRISPR associated (Cas) systems. Viruses *4*, 2291–2311.

Robertson, E.J., Evans, M.J., and Kaufman, M.H. (1983). X-chromosome instability in pluripotential stem cell lines derived from parthenogenetic embryos. Journal of embryology and experimental morphology *74*, 297–309.

Romagnuolo, R., Masoudpour, H., Porta-Sánchez, A., Qiang, B., Barry, J., Laskary, A., Qi, X., Massé, S., Magtibay, K., Kawajiri, H., Wu, J., Valdman Sadikov, T., Rothberg, J., Panchalingam, K.M., Titus, E., Li, R.-K., Zandstra, P.W., Wright, G.A., Nanthakumar, K., Ghugre, N.R., Keller, G., and Laflamme, M.A.

7 REFERENCES

(2019). Human Embryonic Stem Cell-Derived Cardiomyocytes Regenerate the Infarcted Pig Heart but Induce Ventricular Tachyarrhythmias. Stem cell reports *12*, 967–981.

Ronvelia, D., Greenwood, J., Platt, J., Hakim, S., and Zaragoza, M.V. (2012). Intrafamilial variability for novel TAZ gene mutation: Barth syndrome with dilated cardiomyopathy and heart failure in an infant and left ventricular noncompaction in his great-uncle. Molecular genetics and metabolism *107*, 428–432.

Rosenbaum, A.N., Agre, K.E., and Pereira, N.L. (2020). Genetics of dilated cardiomyopathy: practical implications for heart failure management. Nature reviews. Cardiology *17*, 286–297.

Roux-Buisson, N., Cacheux, M., Fourest-Lieuvin, A., Fauconnier, J., Brocard, J., Denjoy, I., Durand, P., Guicheney, P., Kyndt, F., Leenhardt, A., Le Marec, H., Lucet, V., Mabo, P., Probst, V., Monnier, N., Ray, P.F., Santoni, E., Trémeaux, P., Lacampagne, A., Fauré, J., Lunardi, J., and Marty, I. (2012). Absence of triadin, a protein of the calcium release complex, is responsible for cardiac arrhythmia with sudden death in human. Human molecular genetics *21*, 2759–2767.

Sander, J.D., and Joung, J.K. (2014). CRISPR-Cas systems for editing, regulating and targeting genomes. Nature biotechnology *32*, 347–355.

Schweizer, P.A., Darche, F.F., Ullrich, N.D., Geschwill, P., Greber, B., Rivinius, R., Seyler, C., Müller-Decker, K., Draguhn, A., Utikal, J., Koenen, M., Katus, H.A., and Thomas, D. (2017). Subtype-specific differentiation of cardiac pacemaker cell clusters from human induced pluripotent stem cells. Stem cell research & therapy *8*, 229.

Sedaghat-Hamedani, F., Haas, J., Zhu, F., Geier, C., Kayvanpour, E., Liss, M., Lai, A., Frese, K., Pribe-Wolferts, R., Amr, A., Li, D.T., Samani, O.S., Carstensen, A., Bordalo, D.M., Müller, M., Fischer, C., Shao, J., Wang, J., Nie, M., Yuan, L., Haßfeld, S., Schwartz, C., Zhou, M., Zhou, Z., Shu, Y., Wang, M., Huang, K., Zeng, Q., Cheng, L., Fehlmann, T., Ehlermann, P., Keller, A., Dieterich, C., Streckfuß-Bömeke, K., Liao, Y., Gotthardt, M., Katus, H.A., and Meder, B. (2017). Clinical genetics and outcome of left ventricular non-compaction cardiomyopathy. European heart journal *38*, 3449–3460.

Sen, L., Cui, G., Fonarow, G.C., and Laks, H. (2000). Differences in mechanisms of SR dysfunction in ischemic vs. idiopathic dilated cardiomyopathy. American journal of physiology. Heart and circulatory physiology *279*, H709-18.

Shen, H., and Green, M.R. (2006). RS domains contact splicing signals and promote splicing by a common mechanism in yeast through humans. Genes & development *20*, 1755–1765.

Shi, Y., Inoue, H., Wu, J.C., and Yamanaka, S. (2017). Induced pluripotent stem cell technology: a decade of progress. Nature reviews. Drug discovery *16*, 115–130.

Shiba, Y., Gomibuchi, T., Seto, T., Wada, Y., Ichimura, H., Tanaka, Y., Ogasawara, T., Okada, K., Shiba, N., Sakamoto, K., Ido, D., Shiina, T., Ohkura, M., Nakai, J., Uno, N., Kazuki, Y., Oshimura, M., Minami, I., and Ikeda, U. (2016). Allogeneic transplantation of iPS cell-derived cardiomyocytes regenerates primate hearts. Nature *538*, 388–391.

7 REFERENCES

Shinkuma, S., Guo, Z., and Christiano, A.M. (2016). Site-specific genome editing for correction of induced pluripotent stem cells derived from dominant dystrophic epidermolysis bullosa. Proceedings of the National Academy of Sciences of the United States of America *113*, 5676–5681.

Smithies, O., Gregg, R.G., Boggs, S.S., Koralewski, M.A., and Kucherlapati, R.S. (1985). Insertion of DNA sequences into the human chromosomal beta-globin locus by homologous recombination. Nature *317*, 230–234.

Solis, A.S., Shariat, N., and Patton, J.G. (2008). Splicing fidelity, enhancers, and disease. Frontiers in bioscience a journal and virtual library *13*, 1926–1942.

Soto-Velasquez, M., Hayes, M.P., Alpsoy, A., Dykhuizen, E.C., and Watts, V.J. (2018). A Novel CRISPR/Cas9-Based Cellular Model to Explore Adenylyl Cyclase and cAMP Signaling. Molecular pharmacology *94*, 963–972.

Srivastava, D. (2006). Making or breaking the heart: from lineage determination to morphogenesis. Cell *126*, 1037–1048.

Staerk, J., Dawlaty, M.M., Gao, Q., Maetzel, D., Hanna, J., Sommer, C.A., Mostoslavsky, G., and Jaenisch, R. (2010). Reprogramming of human peripheral blood cells to induced pluripotent stem cells. Cell stem cell *7*, 20–24.

Streckfuss-Bömeke, K., Tiburcy, M., Fomin, A., Luo, X., Li, W., Fischer, C., Özcelik, C., Perrot, A., Sossalla, S., Haas, J., Vidal, R.O., Rebs, S., Khadjeh, S., Meder, B., Bonn, S., Linke, W.A., Zimmermann, W.-H., Hasenfuss, G., and Guan, K. (2017). Severe DCM phenotype of patient harboring RBM20 mutation S635A can be modeled by patient-specific induced pluripotent stem cell-derived cardiomyocytes. Journal of molecular and cellular cardiology *113*, 9–21.

Streckfuss-Bömeke, K., Wolf, F., Azizian, A., Stauske, M., Tiburcy, M., Wagner, S., Hübscher, D., Dressel, R., Chen, S., Jende, J., Wulf, G., Lorenz, V., Schön, M.P., Maier, L.S., Zimmermann, W.H., Hasenfuss, G., and Guan, K. (2013). Comparative study of human-induced pluripotent stem cells derived from bone marrow cells, hair keratinocytes, and skin fibroblasts. European heart journal *34*, 2618–2629.

Sun, Q., Guo, J., Hao, C., Guo, R., Hu, X., Chen, Y., Yang, W., Li, W., and Feng, Y. (2020). Whole-exome sequencing reveals two de novo variants in the RBM20 gene in two Chinese patients with left ventricular non-compaction cardiomyopathy. PEDIATRIC INVESTIGATION *4*, 11–16.

Swaminathan, P.D., Purohit, A., Hund, T.J., and Anderson, M.E. (2012). Calmodulin-dependent protein kinase II: linking heart failure and arrhythmias. Circulation research *110*, 1661–1677.

Tajsharghi, H. (2008). Thick and thin filament gene mutations in striated muscle diseases. International journal of molecular sciences *9*, 1259–1275.

Takahashi, K., Tanabe, K., Ohnuki, M., Narita, M., Ichisaka, T., Tomoda, K., and Yamanaka, S. (2007). Induction of pluripotent stem cells from adult human fibroblasts by defined factors. Cell *131*, 861–872.

7 REFERENCES

Takahashi, K., and Yamanaka, S. (2006). Induction of pluripotent stem cells from mouse embryonic and adult fibroblast cultures by defined factors. Cell *126*, 663–676.

Tang, T., Gao, M.H., Lai, N.C., Firth, A.L., Takahashi, T., Guo, T., Yuan, J.X.-J., Roth, D.M., and Hammond, H.K. (2008). Adenylyl cyclase type 6 deletion decreases left ventricular function via impaired calcium handling. Circulation *117*, 61–69.

Tester, D.J., and Ackerman, M.J. (2008). Novel gene and mutation discovery in congenital long QT syndrome: let's keep looking where the street lamp standeth. Heart rhythm *5*, 1282–1284.

Thomson, J.A., Itskovitz-Eldor, J., Shapiro, S.S., Waknitz, M.A., Swiergiel, J.J., Marshall, V.S., and Jones, J.M. (1998). Embryonic stem cell lines derived from human blastocysts. Science (New York, N.Y.) *282*, 1145–1147.

Tiburcy, M., Hudson, J.E., Balfanz, P., Schlick, S., Meyer, T., Chang Liao, M.-L., Levent, E., Raad, F., Zeidler, S., Wingender, E., Riegler, J., Wang, M., Gold, J.D., Kehat, I., Wettwer, E., Ravens, U., Dierickx, P., van Laake, L.W., Goumans, M.J., Khadjeh, S., Toischer, K., Hasenfuss, G., Couture, L.A., Unger, A., Linke, W.A., Araki, T., Neel, B., Keller, G., Gepstein, L., Wu, J.C., and Zimmermann, W.-H. (2017). Defined Engineered Human Myocardium With Advanced Maturation for Applications in Heart Failure Modeling and Repair. Circulation *135*, 1832–1847.

Tiemann, U., Sgodda, M., Warlich, E., Ballmaier, M., Schöler, H.R., Schambach, A., and Cantz, T. (2011). Optimal reprogramming factor stoichiometry increases colony numbers and affects molecular characteristics of murine induced pluripotent stem cells. Cytometry. Part A the journal of the International Society for Analytical Cytology *79*, 426–435.

Tohyama, S., Hattori, F., Sano, M., Hishiki, T., Nagahata, Y., Matsuura, T., Hashimoto, H., Suzuki, T., Yamashita, H., Satoh, Y., Egashira, T., Seki, T., Muraoka, N., Yamakawa, H., Ohgino, Y., Tanaka, T., Yoichi, M., Yuasa, S., Murata, M., Suematsu, M., and Fukuda, K. (2013). Distinct metabolic flow enables large-scale purification of mouse and human pluripotent stem cell-derived cardiomyocytes. Cell stem cell *12*, 127–137.

Turunen, J.J., Niemelä, E.H., Verma, B., and Frilander, M.J. (2013). The significant other: splicing by the minor spliceosome. Wiley interdisciplinary reviews. RNA *4*, 61–76.

Ueno, S., Weidinger, G., Osugi, T., Kohn, A.D., Golob, J.L., Pabon, L., Reinecke, H., Moon, R.T., and Murry, C.E. (2007). Biphasic role for Wnt/beta-catenin signaling in cardiac specification in zebrafish and embryonic stem cells. Proceedings of the National Academy of Sciences of the United States of America *104*, 9685–9690.

van den Hoogenhof, M.M.G., Beqqali, A., Amin, A.S., van der Made, I., Aufiero, S., Khan, M.A.F., Schumacher, C.A., Jansweijer, J.A., van Spaendonck-Zwarts, K.Y., Remme, C.A., Backs, J., Verkerk, A.O., Baartscheer, A., Pinto, Y.M., and Creemers, E.E. (2018). RBM20 Mutations Induce an Arrhythmogenic Dilated Cardiomyopathy Related to Disturbed Calcium Handling. Circulation *138*, 1330–1342.

van der Velden, J., and Stienen, G.J.M. (2019). Cardiac Disorders and Pathophysiology of Sarcomeric Proteins. Physiological reviews *99*, 381–426.

7 REFERENCES

Vandenberk, B., Vandael, E., Robyns, T., Vandenberghe, J., Garweg, C., Foulon, V., Ector, J., and Willems, R. (2016). Which QT Correction Formulae to Use for QT Monitoring? Journal of the American Heart Association: Cardiovascular and Cerebrovascular Disease *5*.

Velychko, S., Adachi, K., Kim, K.-P., Hou, Y., MacCarthy, C.M., Wu, G., and Schöler, H.R. (2019). Excluding Oct4 from Yamanaka Cocktail Unleashes the Developmental Potential of iPSCs. Cell stem cell *25*, 737-753.e4.

Veres, A., Gosis, B.S., Ding, Q., Collins, R., Ragavendran, A., Brand, H., Erdin, S., Cowan, C.A., Talkowski, M.E., and Musunuru, K. (2014). Low incidence of off-target mutations in individual CRISPR-Cas9 and TALEN targeted human stem cell clones detected by whole-genome sequencing. Cell stem cell *15*, 27–30.

Vikhorev, P.G., and Vikhoreva, N.N. (2018). Cardiomyopathies and Related Changes in Contractility of Human Heart Muscle. International journal of molecular sciences *19*.

Wang, Z., and Burge, C.B. (2008). Splicing regulation: from a parts list of regulatory elements to an integrated splicing code. RNA *14*, 802–813.

Warren, C.M., Krzesinski, P.R., Campbell, K.S., Moss, R.L., and Greaser, M.L. (2004). Titin isoform changes in rat myocardium during development. Mechanisms of development *121*, 1301–1312.

Warren, L., Manos, P.D., Ahfeldt, T., Loh, Y.-H., Li, H., Lau, F., Ebina, W., Mandal, P.K., Smith, Z.D., Meissner, A., Daley, G.Q., Brack, A.S., Collins, J.J., Cowan, C., Schlaeger, T.M., and Rossi, D.J. (2010). Highly efficient reprogramming to pluripotency and directed differentiation of human cells with synthetic modified mRNA. Cell stem cell *7*, 618–630.

Watanabe, T., Kimura, A., and Kuroyanagi, H. (2018). Alternative Splicing Regulator RBM20 and Cardiomyopathy. Frontiers in molecular biosciences *5*, 105.

Wengrofsky, P., Armenia, C., Oleszak, F., Kupferstein, E., Rednam, C., Mitre, C.A., and McFarlane, S.I. (2019). Left Ventricular Trabeculation and Noncompaction Cardiomyopathy: A Review. EC clinical and experimental anatomy *2*, 267–283.

Will, C.L., and Lührmann, R. (2011). Spliceosome structure and function. Cold Spring Harbor perspectives in biology *3*.

Wu, H., Yang, H., Rhee, J.-W., Zhang, J.Z., Lam, C.K., Sallam, K., Chang, A.C.Y., Ma, N., Lee, J., Zhang, H., Blau, H.M., Bers, D.M., and Wu, J.C. (2019). Modelling diastolic dysfunction in induced pluripotent stem cell-derived cardiomyocytes from hypertrophic cardiomyopathy patients. European heart journal *40*, 3685–3695.

Wyles, S.P., Li, X., Hrstka, S.C., Reyes, S., Oommen, S., Beraldi, R., Edwards, J., Terzic, A., Olson, T.M., and Nelson, T.J. (2016). Modeling structural and functional deficiencies of RBM20 familial dilated cardiomyopathy using human induced pluripotent stem cells. Human molecular genetics *25*, 254–265.

7 REFERENCES

Xiang, F., Sakata, Y., Cui, L., Youngblood, J.M., Nakagami, H., Liao, J.K., Liao, R., and Chin, M.T. (2006). Transcription factor CHF1/Hey2 suppresses cardiac hypertrophy through an inhibitory interaction with GATA4. American journal of physiology. Heart and circulatory physiology *290*, H1997-2006.

Xu, H., Xiao, T., Chen, C.-H., Li, W., Meyer, C.A., Wu, Q., Di Wu, Le Cong, Zhang, F., Liu, J.S., Brown, M., and Liu, X.S. (2015). Sequence determinants of improved CRISPR sgRNA design. Genome research *25*, 1147–1157.

Xu, X., Gao, D., Wang, P., Chen, J., Ruan, J., Xu, J., and Xia, X. (2018). Efficient homology-directed gene editing by CRISPR/Cas9 in human stem and primary cells using tube electroporation. Scientific reports *8*, 11649.

Xu, X., Yang, D., Ding, J.-H., Wang, W., Chu, P.-H., Dalton, N.D., Wang, H.-Y., Bermingham, J.R., Ye, Z., Liu, F., Rosenfeld, M.G., Manley, J.L., Ross, J., Chen, J., Xiao, R.-P., Cheng, H., and Fu, X.-D. (2005). ASF/SF2-regulated CaMKIIdelta alternative splicing temporally reprograms excitation-contraction coupling in cardiac muscle. Cell *120*, 59–72.

Yang, X., Pabon, L., and Murry, C.E. (2014a). Engineering adolescence: maturation of human pluripotent stem cell-derived cardiomyocytes. Circulation research *114*, 511–523.

Yang, X., Rodriguez, M., Pabon, L., Fischer, K.A., Reinecke, H., Regnier, M., Sniadecki, N.J., Ruohola-Baker, H., and Murry, C.E. (2014b). Tri-iodo-l-thyronine promotes the maturation of human cardiomyocytes-derived from induced pluripotent stem cells. Journal of molecular and cellular cardiology *72*, 296–304.

Yeakley, J.M., Tronchère, H., Olesen, J., Dyck, J.A., Wang, H.Y., and Fu, X.D. (1999). Phosphorylation regulates in vivo interaction and molecular targeting of serine/arginine-rich pre-mRNA splicing factors. The Journal of cell biology *145*, 447–455.

Yoshida, S., Miyagawa, S., Fukushima, S., Kawamura, T., Kashiyama, N., Ohashi, F., Toyofuku, T., Toda, K., and Sawa, Y. (2018). Maturation of Human Induced Pluripotent Stem Cell-Derived Cardiomyocytes by Soluble Factors from Human Mesenchymal Stem Cells. Molecular therapy the journal of the American Society of Gene Therapy *26*, 2681–2695.

Young, C.S., Hicks, M.R., Ermolova, N.V., Nakano, H., Jan, M., Younesi, S., Karumbayaram, S., Kumagai-Cresse, C., Wang, D., Zack, J.A., Kohn, D.B., Nakano, A., Nelson, S.F., Miceli, M.C., Spencer, M.J., and Pyle, A.D. (2016). A Single CRISPR-Cas9 Deletion Strategy that Targets the Majority of DMD Patients Restores Dystrophin Function in hiPSC-Derived Muscle Cells. Cell stem cell *18*, 533–540.

Yu, J., Vodyanik, M.A., Smuga-Otto, K., Antosiewicz-Bourget, J., Frane, J.L., Tian, S., Nie, J., Jonsdottir, G.A., Ruotti, V., Stewart, R., Slukvin, I.I., and Thomson, J.A. (2007). Induced pluripotent stem cell lines derived from human somatic cells. Science (New York, N.Y.) *318*, 1917–1920.

Yue, X.-S., Fujishiro, M., Nishioka, C., Arai, T., Takahashi, E., Gong, J.-S., Akaike, T., and Ito, Y. (2012). Feeder cells support the culture of induced pluripotent stem cells even after chemical fixation. PloS one *7*, e32707.

7 REFERENCES

Zhang, H., Tian, L., Shen, M., Tu, C., Wu, H., Gu, M., Paik, D.T., and Wu, J.C. (2019). Generation of Quiescent Cardiac Fibroblasts From Human Induced Pluripotent Stem Cells for In Vitro Modeling of Cardiac Fibrosis. Circulation research *125*, 552–566.

Zhang, M.Q. (1998). Statistical features of human exons and their flanking regions. Human molecular genetics *7*, 919–932.

Zhang, T., Johnson, E.N., Gu, Y., Morissette, M.R., Sah, V.P., Gigena, M.S., Belke, D.D., Dillmann, W.H., Rogers, T.B., Schulman, H., Ross, J., and Brown, J.H. (2002). The cardiac-specific nuclear delta(B) isoform of Ca2+/calmodulin-dependent protein kinase II induces hypertrophy and dilated cardiomyopathy associated with increased protein phosphatase 2A activity. The Journal of biological chemistry *277*, 1261–1267.

Zheng, T., Hou, Y., Zhang, P., Zhang, Z., Xu, Y., Zhang, L., Niu, L., Yang, Y., Liang, D., Yi, F., Peng, W., Feng, W., Yang, Y., Chen, J., Zhu, Y.Y., Zhang, L.-H., and Du, Q. (2017). Profiling single-guide RNA specificity reveals a mismatch sensitive core sequence. Scientific reports *7*, 40638.

Zhou, W., and Freed, C.R. (2009). Adenoviral gene delivery can reprogram human fibroblasts to induced pluripotent stem cells. Stem cells (Dayton, Ohio) *27*, 2667–2674.

Zhou, Z., and Fu, X.-D. (2013). Regulation of splicing by SR proteins and SR protein-specific kinases. Chromosoma *122*, 191–207.

Zuris, J.A., Thompson, D.B., Shu, Y., Guilinger, J.P., Bessen, J.L., Hu, J.H., Maeder, M.L., Joung, J.K., Chen, Z.-Y., and Liu, D.R. (2015). Cationic lipid-mediated delivery of proteins enables efficient protein-based genome editing in vitro and in vivo. Nature biotechnology *33*, 73–80.